P9-EDD-382

Manufacturing 2000

William L. Duncan

American Management Association

New York • Atlanta • Boston • Chicago • Kansas City • San Francisco • Washington, D.C.
Brussels • Mexico City • Tokyo • Toronto

This book is available at a special discount when ordered in bulk quantities. For information, contact Special Sales Department, AMACOM, a division of American Management Association, 135 West 50th Street, New York, NY 10020.

Library of Congress Cataloging-in-Publication Data

Duncan, William L.
Manufacturing 2000 / William L. Duncan.
p. cm.
Includes bibliographical references and index.
ISBN 0-8144-0235-6
1. Manufactures—Management—Forecasting. I. Title. II. Title: Manufacturing two thousand.
HD9720.5.D86 1994
670'.68—dc20 *94-13528*
CIP

Printing number

10 9 8 7 6 5 4 3 2 1

In the Epistle of James, we learn that

> *"Every good gift, and every perfect gift is from above, and cometh down from the Father. . . ."*

This book is dedicated to the best earthly gifts that God has given to me—my children Erica and Adam. May God grant them as much happiness in their lives ahead as He has brought to me through them.

Contents

Acknowledgments vii

Introduction ix

Chapter 1 The Model 1

Book One Where We've Been and Where We Are 5

Chapter 2 Dominant Themes: Control, Computers, and a Global Focus 7

Chapter 3 Environmental Factors: From Deregulation to Quantum Physics 13

Chapter 4 Dominant Manpower Issues: The End of the Direct Labor Era 26

Chapter 5 Dominant Methods: Doing It by the Numbers 33

Chapter 6 Dominant Materials in the Information Age 61

Chapter 7 Dominant Machines: The Reign of the Computer 77

Chapter 8 Back to the Model 99

Book Two: Where We Are Headed 107

Chapter 9 Dominant Themes: From the 1990s Through the Twenty-First Century 109

Chapter 10 Environmental Factors: Building Alliances in a Fragmented World 113

Chapter 11 Dominant Manpower Issues: Education or Extinction 127

Chapter 12 Dominant Methods: Forging Strong Links in the Global Chain 153

Chapter 13 Dominant Materials: From Software and Silicon to Composites and Ceramics 203

Chapter 14 Dominant Machines: Equipment With a High IQ 234

Chapter 15 Using the Model to Identify Your "Critical Path" and Manage Change in the Twenty-First Century 253

Chapter 16 Conclusions and Recommendations 262

Epilogue: A Personal Note From the Author 267

Bibliography 269

Index 273

Acknowledgments

There are three primary groups of people to whom I am indebted for enabling me to write this book. The first is my army of reviewers including John Kramer, Janet Riley, Vern Pechta, and Jim Duncan (my dad). They did a terrific job and gave me some invaluable advice. I even followed some of it! I owe a special debt of gratitude to Tony Vlamis, who believed in this project and nurtured it through publication.

The second is my family, especially my wife, Linda. Like everyone else who does one of these projects on his own time, I've spent a lot of evenings and weekends over the last few years researching and writing it. It was a labor of love for me, and a great deal of me went into it that would otherwise have been given to my family.

Third, there is a large group of people who have been quoted in this book, referred to in this book, or whose previous works have helped to shape my thinking on this subject. People like Gilbreath, Cetron, Hamrin, Thurow, and Drexler. Giants in the business of Future Studies. I have done my best to give every quoted and referenced individual proper attribution in this book. If I've missed the mark, either in wording or context, I humbly apologize. I also found during the course of reviewing this manuscript that I've quoted several *Fortune* magazine articles on various topics. Time after time I found the writing and insight of the work in *Fortune* to be so poignant, so "right on target," that I just couldn't resist quoting it. I don't know where *Fortune* finds its writers, but I have grown to have a lot of respect for what they say.

Although I didn't interview them specifically for this book, I've also been profoundly influenced by the professionals I've encountered in American manufacturing over the last twenty-plus years. They are the ones who don't "do it for the money," or for power, or for prestige. They are committed to be exactly what they are—the best there is at what they do. Ultimately, this book is written for them, for the rest of you just like them, and, perhaps most of all, for all of our children.

Introduction

One hot summer day in 1989, I sat listening to the vice president in charge of facilities at a major defense contractor as he described a plant he'd just finished touring in the Midwest. His narrative was appropriately framed by the occasional roar of an F-15 Eagle lifting off from the airfield for a test flight. I leaned closer over the conference room table to overcome the noisy background, then leaned back again as it faded, oblivious to it because I was so caught up in his description.

He recounted a facility that was virtually perfect for the manufacture of aircraft—a large, open, high-bay structure designed to accommodate several of the most desirable aspects of leading-edge manufacturing, including adjacent subassembly and final assembly operations, point-of-use parts delivery, and continuous material flow.

All of the meeting attendees were impressed even before he came to what for me was the most poignant observation of all: The facility he had been describing was built—for just this purpose—in 1939!

"What did they know fifty years ago that enabled these people to design a manufacturing facility so uniquely suited to leading-edge production techniques in 1989?" he asked.

Somehow, the management of this aircraft manufacturer had achieved something extremely rare: a genuine vision of their own future. They were able to project technological, product, and process evolution clearly enough to attain an extraordinary success in facility design.

Today, this kind of visionary talent is far more crucial than it was fifty years ago. It is no longer sufficient to react (even quickly) to change. In order to be successful, managers will be required to anticipate change, initiate change, and shape change.

Change management became an increasingly popular topic in management literature through the late 1980s, and this was entirely appropriate. Never has the number of changes or the pervasiveness of change been so strongly felt. The pace of technology development is rapid and accelerating. The labor force is shrinking and highly mobile. Our markets and competitors are increasingly global. The protection of our environment has become an important factor in our day-to-day decisions. Consider also government regulation and deregulation, the dynamic energy picture, the evolution of workers' values, and the critical need for education reform. These changes and others have created a volatile and enigmatic backdrop against which to develop successful business strategies.

The purpose of this book is to present a predictive and analytical model that I and my clients have found to be very useful in recent years. The model provides a framework to identify and evaluate trends in technology and the manufacturing environment, allowing us to create a vision of our organizations' future. Beyond this, it provides us an opportunity to orchestrate the actions required to achieve that vision, the activities that constitute our *critical path*.

I will cover each aspect of the model in detail and use practical examples as much as possible to illustrate the model's use. When I have completed the examination of the model, I will discuss some individual findings and conclusions based upon its use, categorizing these by industry, company size, and so forth as appropriate.

Finally, I will include some general recommendations based on these conclusions. However, the recommendations and even the findings and conclusions are only as good as the author's ability to interpret the model data. They are of secondary importance at best.

When going through the first half of this book, the reader should ask him- or herself the following question: "Is my company at least this far along in its development?" The book should be useful as a benchmark to compare an individual company to current industry practices.

Reading the second half of this book will provide useful insights about what is likely to transpire over the decades ahead in each area of the company. Strategic planners and operations planners are likely to spend most of their time in this portion of the book.

Everyone involved in manufacturing will find things of interest in this text. However, the real value of this book lies not in the author's perceptions, but in the reader's ability to understand and utilize the model to do what one of my friends refers to as "thinking in three dimensions"; to literally lay out in time the activities that will be critical to the success of his or her own organization.

Chapter 1

The Model

Why Do We Need a Model?

Given the tremendous complexity and interdependence of current trends in technology and manufacturing, how can we hope to analyze their implications for our individual organizations over the coming decades? How can we attain the kind of vision that produced the aircraft manufacturing facility described in the Introduction? I have found the manufacturing model depicted in Exhibit 1-1 to be a powerful tool for developing this vision. Properly applied, it provides an invaluable structure for analyzing and incorporating projected changes in technology, methodology, manpower, and materials. It can be used to address individual processes, divisions, companies, and entire industries.

All models have limitations, however, and this one is no exception. Robert Gilbreath, in his book *Forward Thinking* (McGraw-Hill, 1987), states:

> The search for a perfect model, then, does not result in the most complex or "realistic" version but the most pragmatic one: the one that balances simplicity with accuracy, cost with benefit.
>
> Whether crude or exotic, primitive or technologically advanced, manual or automated, belonging solely in one person's mind or owned and operated by scores of managers in unison, each model is limited, contrived, and of no intrinsic value. Its only value is derived from the benefits it brings its users. Business models are tools, not ends in themselves, and the nature of change is that it mocks most of them.

The model shown in Exhibit 1-1 is deceptively simple in its structure. The elements of manufacturing, which some readers will recognize from the Ishikawa "fishbone" diagram, are listed along the y-axis. These elements are manpower, material, methods, and machines. Along the x-axis are aligned the fundamental business processes: business administration, obtaining business, defining products and processes, production, distribution, and product/customer support. The z-axis represents time. The arrow shown pointing toward the solids model illustrates the pervasive and ever-present impact of environmental forces on the development of manufacturing.

When one performs a three-dimensional cross-impact analysis using this model, myriad relationships become visible that can demonstrate potential market opportunities, areas of needed R&D, training and education requirements, manpower needs, and appropriate timing. Incorporating specific developments anticipated in each cell of the matrix allows one to construct a vision of an individual organization or an entire industry.

Exhibit 1-1. The solids model of manufacturing.

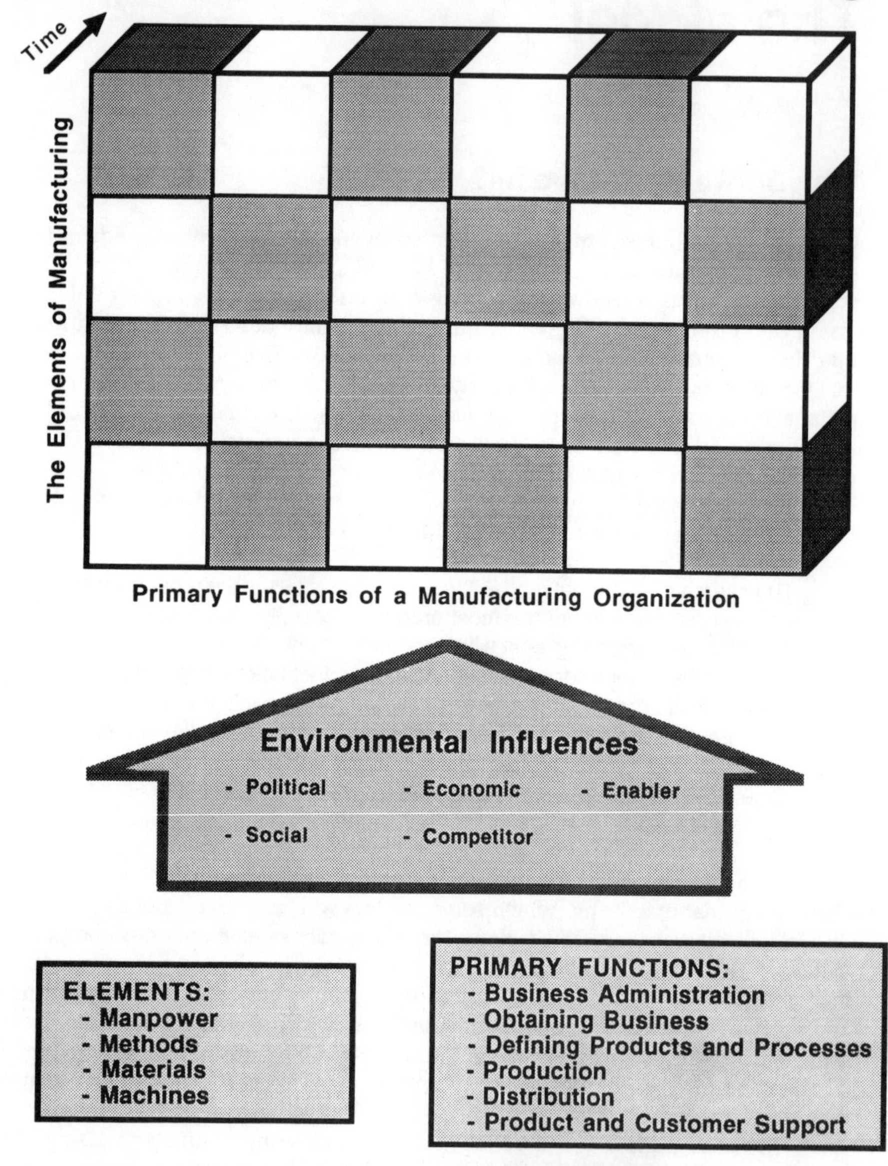

Exhibit 1-2. Years between major milestones in manufacturing history.

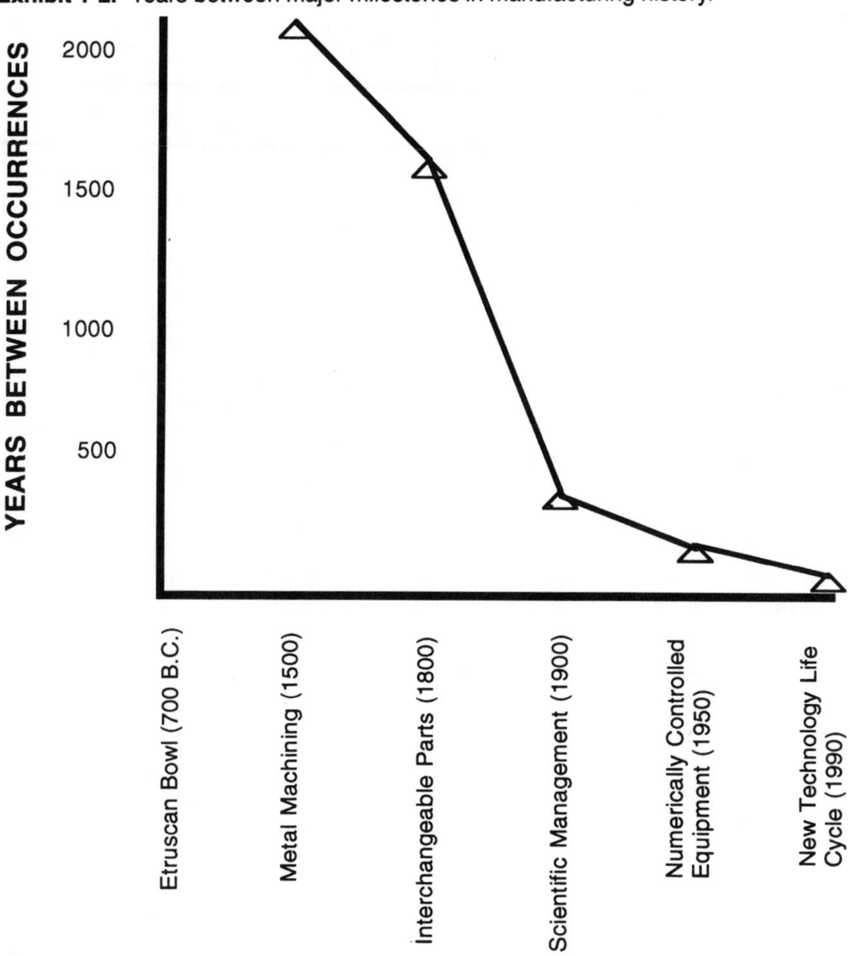

In subsequent chapters, we'll review our progress and current position as American manufacturers in each plane of the model, focusing on recent developments (that is, major developments and trends since the 1960s). Then we will reflect on likely future developments in each sector and try to identify some instances of convergence. Before we examine these areas, however, it is useful to review the foundation upon which today's manufacturing is built.

From Etruscan Bowls to Biotechnology

The origins of manufacturing are said to lie in the production of weapons and dishes with crude tools several hundred years before Christ. Evidence of this activity has been found in Etruscan wooden bowl fragments from the Tomb of the

Warrior at Corneto, dating from 700 B.C. Metal machining appeared in the fifteenth century and really took root in the eighteenth century with the advent of early machine tools in production settings. The next major step in the evolution of manufacturing was probably the development of interchangeable parts, which occurred as a result of Eli Whitney's work with firearms around 1800. With that development came the understanding of economies of scale and thus the concept of mass production. The first formal recognition of the benefits of applying scientific method analysis to both people and machines came from work done by Frederick W. Taylor, the father of manufacturing "standards," in the first decade of the 1900s. The next step of major significance occurred in the 1950s with the advent of initial versions of numerically controlled machines.

More recent developments are covered in depth within later chapters; however, it might be interesting now to note the time compression curve depicted in Exhibit 1-2. The time between creation of Etruscan bowls and metal machining was roughly 2,100 years. The span from metal machining to interchangeable parts and mass production concept development covered 300 years. The time between mass production and scientific work measurement and standards was 100 years. The period between Taylor's work and the first numerically controlled equipment spanned only 50 years. Anyone involved with the fields of computer technology, biotechnology, or materials engineering can attest that today the life cycle for generations of new technologies is running less than ten years. And so it goes, as we hurtle at blinding speed into the future of manufacturing.

BOOK ONE:

WHERE WE'VE BEEN AND WHERE WE ARE

Chapter 2

Dominant Themes: Control, Computers, and a Global Focus

The 1960s: Riding the Crest of the Wave

Mainstream American manufacturers since the 1960s have undergone major changes in the way they do business (see Exhibit 2-1).

In the 1960s, a great deal of emphasis was placed on what is referred to as basic controls. Basic controls in this context involve the development of accurate bills of material, accurate inventory accounting, accurate production routings, and consistent structures for costing products and processes.

These activities were necessary because we were "catching our breath" as a nation. We were riding the crest of one of the world's most productive eras.

Exhibit 2-1. Dominant themes in American manufacturing.

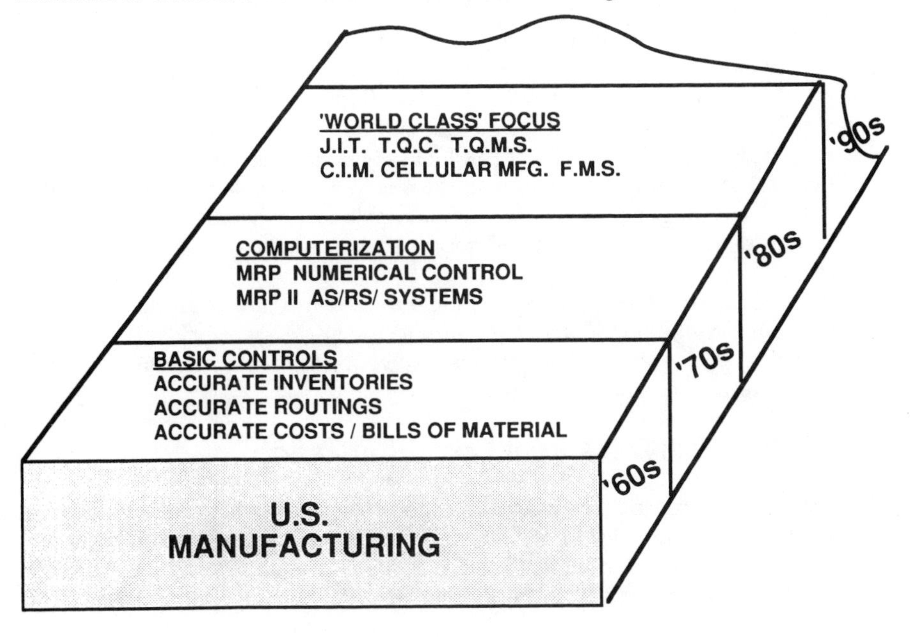

Between 1950 and 1968, our productivity growth dipped below the rate of 2 percent annually only three times.

In contrast, the 2 percent rate was *exceeded* only four times between 1969 and 1986. Between 1986 and 1991, we grew in terms of productivity only another 3 percent in total.) We were so incredibly busy satisfying the unflagging demand for our products during the 1950s and early 1960s that we were able to take advantage of economies of scale and mass production methods to generate tremendous growth.

Beginning with the postwar period, several countries that began far behind the United States in terms of resources, technology, and skilled workforce also began to make marked gains in productivity. For example, the total productivity gain of the United States between 1950 and 1983 was 129 percent (measured in output per worker hour). During that same period, Canadian productivity rose 215 percent, France gained 458 percent, and West Germany was up 508 percent. Japanese worker productivity was up 1,624 percent. Many analysts credit these gains in our competitors' standings to the dissemination of advanced manufacturing equipment and systems in those countries. However, few foreign companies could compete with either the indigenous talent or the economies of scale that were the hallmark of American firms like General Motors, General Electric, and Westinghouse. Even with the gains of foreign competition, the United States remained dominant throughout the 1960s and well into the 1970s.

Along with rapid growth came rapid loss of control. Inventory levels were of little concern, since raw materials and customers were both in plentiful supply. The cold war era also fostered "progress payments" and a "cost plus" approach on the part of government, so that emphasis in defense contracting environments was sometimes focused on accumulating every possible cost to pass along to the customer (Uncle Sam), rather than on efficiently utilizing resources to minimize waste.

By the middle of the 1960s, many American manufacturers were beginning to come to grips with the need to get tighter control of their operations. This was especially noticeable in the commercial sector. Support staff levels and overhead costs really began to swell, with record accuracy deteriorating and lack of information timeliness contributing to longer and longer manufacturing throughput times. The first real computers were beginning to appear, but they were largely relegated to accounting management tool, the planning cycle was generally monthly at best, hardly an adequate way to respond to dynamic markets and daily production needs.

By the close of the 1960s, computer memory and speed were reaching a critical juncture with computing cost. Computer technology was quickly becoming a practical means for resolving many problems that involve extensive number crunching (repetitive mathematical analysis).

The 1970s: The Decade of Computerization . . .

Two major factors contributed to the widespread deployment of computers within manufacturing in the 1970s: material requirements planning (MRP) and numerical control (NC) of manufacturing machinery.

MRP is a requirements planning system for production operations that accepts a master production schedule for finished products, explodes a documented product structure (bill of material) for the products, nets the individual component and raw material requirements against on-hand and on-order inventory, offsets the remaining requirements by designated procurement or manufacturing lead times, and then recommends order start dates for procurement and component part production.

MRP evolved to encompass more than requirements planning (i.e., purchasing, forecasting, and so on) and hence the moniker also evolved, to MRP II (manufacturing resource planning). Under this banner, mountains of software and hardware have been sold and installed across the United States since 1970, and these systems continue to be implemented today.

Numerical control is a means of operating production machines (most commonly punch presses and machining equipment) by means of numerical instruction. Early applications involved the use of a special punched tape that had to be loaded directly into the machine. Later, with the advent of direct link-ups and programmable logic controllers (PLCs), direct numerical control (DNC) became possible. A single computer could code, store, and download the numerical instructions to multiple machines without the use of tapes and punch readers.

A great deal of progress was made in the decade of the 1970s, and computerization was at least partially responsible. However, an unfortunate side effect also occurred. Management became so enamored with the computer during this period that it was regarded (at least subconsciously) by many as a panacea for all problems. "If we can just get it into the computer," they thought, "we can solve it."

. . . And Out-of-Control Inventories

With the speed and dogged repetitiveness that only a computer can generate, American industry produced too many parts, produced the wrong parts, often at the wrong times, and far too often of very poor quality. We saw inventories swell dramatically during that decade, and our response was quite straightforward. We managed it.

We built enormous high-rise inventory storage facilities, applying state-of-the-art technology to automatically store and retrieve the parts. We coded pages and pages of software to control the automated storage and retrieval systems (AS/RSs) and to track the inventory itself. We created entire departments of people to evaluate and control inventory, develop optimal lot sizes, construct material-handling systems, and determine how to treat the burgeoning inventories to best tax advantage. Predictably (did someone say *predictably*?), overhead costs soared.

A particularly painful realization for many of us toward the end of the 1970s was that MRP/MRP II, while delivering on its promise of being an outstanding (probably as yet unsurpassed) planning tool, was usually lousy when it was used to manage shop floor operations. Its relatively static plans for shop floor production were simply not responsive enough to dynamic factory operations and market shifts.

The 1980s: Made Anywhere But Here

In the 1980s, we came to the abrupt and painful recognition that foreign competitors and foreign markets were eroding our position of preeminence in manufacturing. Specifically in the case of our Japanese competitors, we attributed this situation to the application of just-in-time (JIT) manufacturing and an associated emphasis on product quality.

The problems of manufacturing were masked to some degree by the otherwise still positive growth in U.S. corporate profitability. After accounting for taxes, inflation, and new facilities, corporate profits rose about 22 percent annually between 1982 and 1986. It appears in retrospect that our unparalleled marketing and distribution abilities greatly overshadowed our losses in manufacturing. While wholesale and retail profits rose 150 percent between 1980 and 1986, the profits of primarily manufacturing companies actually declined by 5 percent. Between 1986 and 1992, corporate profits rose an average of only 4.4 percent.

Another important shift began to manifest itself during this period as well. Manufacturing began to gain considerable speed in its transition from smokestack to high-tech industries. Items such as computers, printers, scientific equipment, photographic equipment, and office equipment grew in sales by over 200 percent between 1970 and 1984, while categories such as basic metal products crept up at a rate of about 50 percent in the same period.

Overall, the annual growth rate for high-tech industries between 1977 and 1984 was roughly 14 percent. By the end of 1984, these industries accounted for about 13 percent of U.S. industrial output. Since the 1960s, the share of American manufactured goods represented by high tech has gone from about 25 percent to roughly 50 percent.

However, the middle of the 1980s brought a fierce onslaught of foreign competition. It hit hardest in industries like automobiles, machine tools, and steel production, but high-tech industries were hurt as well. The American steel industry is a vivid example. It was the largest and most modern in the world in the postwar decades. Then between 1975 and 1985 it saw product demand levels dwindle and earnings collapse to near zero. By 1986, imported steel represented 37 percent of domestic consumption.

Another example is American automobiles, which dominated the domestic market through the 1950s and 1960s. By 1990, foreign-built cars represented 35 percent of that same market.

By 1985, even America's high-tech companies were beginning to react to foreign pressure with major cost-cutting efforts. AT&T, IBM, MCI, and Wang initiated unprecedented write-offs, layoffs, and plant closings. Between 1980 and 1986, U.S. exports of high-tech goods rose 29 percent, while imports of high-tech equipment rose about 250 percent. It became clear that foreign competitors, and the Japanese in particular, were beating us at our own game: manufacturing.

As one might expect, different companies reacted in different ways to this situation. Certain response trends could be seen even by specific industry. Almost everything was and still is being tried. Among the most prolific controlled responses were the adoption of the Japanese stockless production/just-in-time/total quality management techniques, and an approach to total computer-based integration commonly referred to as computer integrated manufacturing (CIM).

JIT and the related concept of stockless production are centered on the idea of eliminating waste in all its forms throughout the manufacturing company. *Waste* here is broadly defined and includes anything and everything that does not add value to the product. Inventory, throughput time, management layers, approval signatures, and inspection are all examples of items and activities that merit minimization efforts in this scenario. Because of the severe constraints placed on Japanese companies in terms of resources (e.g., raw materials or usable land), it is easy to see how this philosophy was nurtured and took root. When it was coupled with the discipline and teaming concepts indigenous to that culture, Japan's JIT manufacturing became the epitome of the term *world class*.

The 1990s: Worshiping at the Quality Altar

By the latter half of the 1980s, it became clear that quality was another critical success factor for American manufacturing in a global economy. In 1987, a landmark book was written by Robert D. Buzzell and Bradley T. Gale entitled *The PIMS Principles* (Free Press). Chapter 6 begins with these words:

> The 1960s and 1970s brought a dawning realization that market share is key to a company's growth and profitability. The 1980s have shown just as clearly that one factor above all others—quality—drives market share. And when superior quality and large market share are both present, profitability is virtually guaranteed.

The authors go on to document in startling detail how quality affects market share, profitability, and growth.

Out of this realization, and the overwhelming perception that foreign products were superior in quality to domestic offerings, a movement somewhat akin to a new religion arose in America. Through its evolution, this movement has been known as statistical quality control (SQC), total quality control (TQC), total quality management (TQM), and total quality management systems (TQMS). Its origins hearken back to the late W. Edwards Deming, whose fourteen-point program for quality improvement is often credited as a major factor in the revitalization of postwar Japanese manufacturing.

Deming, an American, became fascinated with relationships he discovered between working conditions and productivity while studying for his Ph.D. in physics at Yale. Following his formal education, he was instrumental in several SQC implementations in U.S. factories. However, as mentioned earlier, postwar America was so caught up in its own growth that there was little time to worry about quality. Demand was insatiable, and there was little perceived need on the part of management to do anything other than produce in mass. Disillusioned by the lack of upper-level management commitment, Deming was ripe for the enormous opportunity that presented itself to him while he was on assignment planning a national census of the population of Japan.

Deming preached his statistical quality control message to a manufacturing industry desperate for help and willing to try something new. The rest is, as they say, history. Quality is now a big business. Deming, Juran, Crosby, and hosts of others consulted, wrote, and seminared their way to public prominence and lucrative incomes. Today, there are scores of professional societies, special interest groups, and educational organizations dedicated to quality in specific industries and in general. Perhaps the largest is the American Society for Quality Control (ASQC), and one of the most prestigious is the group that confers the coveted

Malcolm Baldrige Award upon the companies it believes best exemplify "total quality."

Whereas JIT and TQM are often viewed as an overall shift on the part of management toward simplification, computer integrated manufacturing (CIM) represented in many instances an antithetical alternative to that approach. JIT is generally less reliant on complex management systems and computers in general; CIM is dedicated to the total integration and control of company operations from marketing and design through production and distribution. The CIM concept has been defined repeatedly but ambiguously over the last decade. I enjoyed this particular passage from a paper written by Joe Aiello of Sperry Corporation (published in *APICS* [American Production and Inventory Control Society] *Seminar Proceedings,* April 1985):

> CIM can be so big and so complicated that unless we can define to ourselves what we want to do and accomplish, we may never implement it successfully.

As many executives and line manufacturing people at companies like General Motors could emphatically attest, truer words were never spoken.

Conclusion: This Isn't the Conclusion

Looking back on the last three decades, one could easily get the impression that the power and importance of manufacturing, at least in the United States, are coming to an end. Indeed, this conclusion is easy to support in popular literature by prominent futurists. It doesn't need to be that way, though, and our country will be far poorer in many ways if that prophecy becomes self-fulfilling. Manufacturing is a megaindustry in transition from smokestack industrialization to computerized, precision material conversion and process management. The dominant themes reviewed in this chapter describe the essential foundation blocks that had to be laid to prepare us for that transition: fundamental operating controls, computerization, and quality. Although our foreign competitors have managed to emulate some of these elements and even surpass us in others, none of them have acquired the skills and accumulated the hard-won lessons we have over the last thirty years. This advantage is far more powerful than most pundits recognize.

Chapter 3

Environmental Factors: From Deregulation to Quantum Physics

It is difficult to overstate the importance of environmental factors in American manufacturing. Environmental factors are the least predictable and, unfortunately, often among the most influential factors in the development of companies, industries, and entire countries. Next, we consider a few examples of environmental aspects from the model presented in Chapter 1 (see Exhibit 3-1).

Political Influences

One example of political influences is the impact of the U.S. embargo on grain against the Soviet Union in the mid 1970s. The serious effects of this embargo were felt by U.S. agricultural equipment manufacturers for many years afterward.

Among the most pervasive political impacts on American manufacturing in the last twenty years was the federal government's decision to aggressively pursue deregulation. This decision has been credited by experts with revitalizing such key American industries as telecommunications and transportation by encouraging innovation through increased competition.

Getting Out of Government's Grip

The Interstate Commerce Commission (ICC), Civil Aeronautics Board (CAB), Federal Aviation Administration (FAA), Federal Deposit Insurance Corporation (FDIC), and the Federal Reserve Board (the Fed) were all formed as a result of massive regulatory efforts on the part of our federal government between the 1880s and the 1930s. In the late 1960s and throughout the 1970s, the value of many of these agencies was brought into question, primarily by conservatives. The result of ensuing investigations was a broad initiative to deregulate three primary economic sectors: finance, transportation, and telecommunications. Major steps along this path to date include a 1970 Federal Reserve Board decision to free interest rates on bank deposits exceeding $100,000 with maturities under six months, a 1974 antitrust suit filed by the Justice Department against AT&T, and a 1980 decision by Congress to deregulate trucking and railroads.

The effects of deregulation on manufacturing are not all obvious, but they are significant. Commercial aircraft order backlogs grew dramatically. More than twenty-five new interstate carriers entered the industry after the deregulation act

Exhibit 3-1. Environmental factors affecting American manufacturing.

Environmental Influences Impacting American Manufacturing

Political:
- Deregulation
- Foriegn Trade Restrictions
- R & D Funding Policies
- Corporate Tax Laws

Competitor:
- Consortia
- Disasters
- Strategic Alliances
- Industrial Espionage

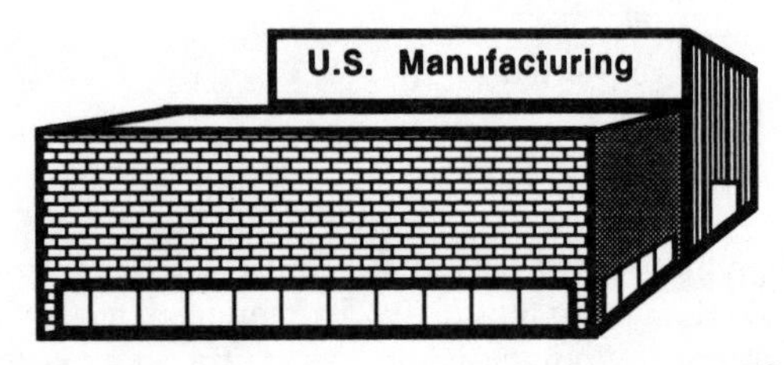

Enablers:
- Quantum Physics
- Biotechnology
- Microprocessors
- Materials Technology

Social:
- Unions
- Value Changes
- Education
- Environmentalism

Economic:
- Minimum Wage Laws
- Interest Rates
- Investment Incentives
- Mergers & Acquisitions
- Unemployment Levels
- National Debt / Budget deficit
- Health Care Costs

was passed by Congress in 1978, in contrast to no new entries during the previous forty years. In addition, travel became more affordable in many locations, with average fares down 13 percent since 1978 (adjusted for inflation). Similarly, passage of the Motor Carrier Act of 1980 resulted in a greatly expanded number of ICC-authorized carriers (up roughly 183 percent between 1980 and 1984), making transportation rates more competitive (some studies indicate a 25 percent reduction in truckload rates between 1977 and 1982) and service better. For transportation, purchasing, and distribution departments of American manufacturers, deregulation offered some real opportunities.

The Foreign Trade Quandary

Over the last ten years, a real conflict has arisen within the United States regarding how to deal with foreign trade. Simply stated, imported goods threaten U.S. industries employing thousands of workers. However, most American consumers are hungry for these goods because they perceive them to be of superior quality at lower prices than domestic counterparts. Trade restrictions levied to stem the influx of imported goods would deprive the consumer of these goods and tend to escalate international friction. Consider this excerpt from an interview with Akio Morita, cofounder of Sony Corporation, published in the September 25, 1989, issue of *Fortune:*

> To buttress his assertion that U.S. industry has itself to blame for much of the trade deficit, Morita recalls a golf game with an American friend in the New York City suburbs. "As I stood on the tee I took out my [U.S. made] Mac Gregor driver. My friend, who likes to tell me that Japan is unfair and how appalling that is, pulled out a [Japanese made] Yonex. I chided him for using these clubs. He replied that his Yonex clubs gave him better purchase on the ball. So we set off on the round, safe in the new understanding that using Japanese clubs was a necessary evil.
>
> After the round he invited me back to his home and showed me around while his wife prepared an evening meal. In the garage were a snowmobile, a motorboat, and a four-wheel-drive vehicle—all Japanese. In the house he had a Sony television and stereo. In fact, there were Japanese products everywhere. So I asked him frankly: "Time and again you have angrily claimed that Japan doesn't buy American products, but if you yourself use only Japanese-made things, what do you suggest we can buy from you?"

Therein lies a real problem for American manufacturers. Imported goods represented 13.4 percent of all goods sold in the United States in 1980, 19 percent in 1986, and 25 percent by 1987. According to *America's New Economy* by Robert Hamrin (Franklin Watts, 1988), 70 percent of the goods manufactured by U.S. firms face severe threats from foreign competition.

As a result of the increasing foreign competition, Congress was presented with literally hundreds of protectionist trade initiatives in the mid-1980s. A major omnibus trade bill was indeed passed in 1987 by both the House and Senate against the wishes of the administration. Over the past several years, more than a third of the market for domestically manufactured goods has been protected by quotas and other trade restrictions.

Conversely, foreign trade barriers are often cited as a real problem for U.S. manufactured exports. The Department of Commerce has stated that if the Japanese eliminated their trade restrictions, American exports would increase by as much as $17 billion. One proposed method for dealing with this problem is to leverage regional trading alliances, combating the EEC and Asian trading blocks through solidarity-building approaches such as the North American Free Trade Agreement (NAFTA). There are strong advantages and disadvantages to this approach, and between Ross Perot and Vice President Gore they have already been clearly defined. With NAFTA now passed and moving ahead, it will still be several years until we know whether the measure works as planned.

An excellent study of U.S. manufacturing by Michael L. Dertouzos and colleagues published in 1989 and entitled *Made in America* (MIT Press) states:

> Our industry reports provide abundant evidence of the impact of government on industrial strategies and outcomes. The collapse of the U.S. consumer-electronics industry was affected in part by tariffs, quotas, antidumping and antitrust laws, and the way those laws were implemented. Similarly, in the textile industry differences in quota categories between American and European tariff regimes explains at least some of the differences in the ways this industry has rebuilt itself in the United States and Western Europe in the face of stiff third-world competition. Many other instances could be cited: environmental and health and safety regulations' influence on the profitability of the automobile industry; the lengthy approval procedures of the U.S. Food and Drug Administration delay the marketing of new drugs; antitrust legislation has been an obstacle to cooperative research in the machine tool and semiconductor industries.

R&D Funding

Government policy on R&D funding and technology development are hotly debated issues and have had a powerful impact on what has been developed and what has not over the last thirty years. The basic research system in the United States remains unsurpassed by any competing nation, even today. However, the organization of governmental entities is poor in the sense that it does not allow them to respond effectively to the need for specific support structures in commercial technology. In other words, we aren't very good at getting basic R&D converted to usable commercial products. Coordination is lacking, as is a clear and singular sense of purpose and priorities in the disbursement of funding. At least a dozen governmental agencies currently have some responsibility for research and development. With fully half of all R&D funding springing from the federal government, this lack of coordination is a real handicap in the face of more focused and coordinated efforts by governmental agencies of competitors, such as Japan's MITI.

Corporate Taxation

Corporate taxes are certainly another factor in the profitability of American manufacturing companies. The "effective tax rate" (the rate of taxation after such adjustments as deductions and credits) for American corporations has declined steadily from over 40 percent in the 1950s to around 13 percent in the early 1980s. However, the effective rates vary dramatically between industries, and even between companies within an industry, depending on how well the tax code is handled by individual companies' accounting departments.

Economic Influences

Economic influences, such as changes in minimum wage levels, interest rates, and tax incentives for capital investment, have also proved to be industry-shaping forces. For example, it is interesting to watch the interplay of legislation already on the books and other laws still in the "proposed" stages with the efforts to promote "free trade." As manufacturers attempt to compete in the near term with other labor markets not constrained by these laws, we will continue to feel great pressures in America to reduce the impact of labor differentials through automation and other general labor efficiencies.

Especially noteworthy are tax laws, which continue to play an important role in providing incentives or disincentives for capital investments in the United States, particularly in those areas where, from year to year, we see changes in what is considered "allowable." Capital intensity per unit manufactured has increased, making this an increasingly important factor in net profitability. As we'll see in subsequent chapters, the importance of this area will likely accelerate even more dramatically.

Merger Mania

Corporate mergers and divestitures are also primary economic forces impacting American manufacturing today. "Merger mania," the accelerating number of mergers, has been matched and exceeded only by the dollar amounts involved. Consider recent history:

Year	*Approximate Number of Mergers*	*Dollar Volume of Mergers*
1984	2,500	$122 billion
1985	3,000	$138 billion
1986	4,000	$190 billion
1990	4,168	$172 billion

Individual merger transaction size has also become a source of concern. Of note are the acquisition of General Foods by Phillip Morris ($5.75 billion), General Motors' acquisition of Hughes Aircraft ($5.2 billion), and R. J. Reynolds' acquisition of Nabisco ($4.9 billion). Mergers and acquisitions often simply don't work. In fact, studies reported by *Business Week* in 1985 stated that over half of them don't work, with a third of them later undone. Despite this data, acquisitions and mergers are increasingly the method of choice for companies attempting to extend their markets overseas. This is particularly telling in light of the recent opening of many European markets, since it is likely to remain a popular strategy for companies moving to that growing market.

In addition, emerging industry companies such as biotechnology businesses reported rate marketing capability and capital availability as paramount reasons for considering consolidation through mergers and acquisitions in the current environment. Reflecting on the recent experiences in the electronics, semiconductor, and auto industries, one can only surmise that merger mania is destined to remain a powerful economic force for some time.

Unemployment's Toll

On a decade-by-decade basis, unemployment has risen steadily from 4.5 percent in the 1950s to 4.8 percent in the 1960s to 6.2 percent in the 1970s and 8.0 percent in the 1980s. (The rates vary greatly across industries, however; rates for service industry occupations are much lower than for blue-collar occupations.) Since the unemployment rate is figured by dividing the number of unemployed persons by the total number of people in the labor force, a good many demographic factors (not the least of which is population itself) come into play. Although unemployment is often considered a by-product of the demise of American manufacturing, it is no less a contributor to that decline, in the sense that it drains the savings of Americans and thereby dries up funding sources for R&D as well as capital expansion.

A few of the more poignant facts involved in the manufacturing and unemployment picture presented in Robert Hamrin's *America's New Economy* (Franklin Watts, 1988) are these:

> From 1970 to 1985, manufacturing's share of U.S. employment declined from 26.1 percent to 19.3 percent. From 1979 to mid-1987, 1.9 million manufacturing jobs were lost. In 1985, manufacturing lost 180,000 jobs while 3 million jobs overall were added to the nonfarm payrolls.
>
> A total of 10.8 million workers 20 years of age and over lost jobs because of plant closings or employment cutbacks over the January 1981–January 1986 period.
>
> Of the 3.4 million workers who found work during the displacement, 2.7 million were working at full-time wage and salary jobs, 56 percent were earning as much or more in their new jobs.

As automation and the general trend toward offshore outsourcing continue, it is unlikely that this situation will improve in the near term. Manufacturing can, and someday will, generate a great number of jobs again. However, they won't be the "touch labor" jobs of the 1950s and 1960s, as we'll see in subsequent chapters.

Deficit, Debt, and the Dollar

The national debt and the budget deficit have also proved to be significant in shaping the American manufacturing environment. There are many interesting ways to view the current level of national debt and reflect on how it came to be so large in so short a time. Relative to the size of the U.S. economy, our debt is larger than that of the majority of other industrial nations. It rose by over 700 percent between 1965 and 1986 alone. In terms of the budget deficit, the situation is most striking because of its abruptness.

Budget deficits were virtually insignificant until 1975. They really began a steep incline around 1982, growing to over 200 billion dollars between 1969 and 1983. Of course, as the deficit increases, so do the interest payments on the debt. By 1986, the government was spending more than half as much on national debt interest payments as it was spending on defense!

The impact of this situation on industry falls into two primary categories. First of all, it reduces the level of capital available for investment, making it more expensive for manufacturers to borrow money for expansion and product development. (Consider the fact that in 1960 roughly 90 percent of the nation's net savings was available for investment; yet by 1984, only about 33 percent was available.) Second, prominent economists believe the budget deficit is a primary cause of increased strength of the U.S. dollar relative to competitors' currencies. This, in turn, drives us toward being a capital importer.

Inflation and Health Care: Taking a Bite out of Profits

Inflation (most commonly measured by changes in the Consumer Price Index, a picture of prices on common consumer goods) is another economic factor that forcefully affects manufacturers. Many of the significant operating costs of manufacturers, including raw materials and energy, are inflation-sensitive. Managing these costs, and often passing them along to the customer, has been a real factor in company profitability. (To grasp the magnitude of the price increases involved, consider that the company car—or your own private car for that matter—cost about $3,200 in 1973. By 1982, it cost about $9,000.)

Inflation in industrialized countries rose dramatically in the 1970s and early 1980s, moderating in the latter half of the 1980s. Typically, a "disinflationary" environment such as the one we experienced in the latter half of the 1980s benefits companies with minimal debt, companies with high productivity rates, and companies with sound technology application ability. Companies most vulnerable in this setting are capital-intensive companies with little pricing flexibility. Specific inflation drivers that have become important to watch include oil prices and our budget deficit (and therefore the value of the dollar).

Health care costs are rising at an alarming rate and have begun to pose serious problems for employers. Technology is improving, which is keeping Americans alive longer and keeping us healthier. But the technology is expensive, and as the baby boomers age, costs are rising drastically, the result of straightforward supply-and-demand economics. According to Marvin Cetron and Owen Davies in *American Renaissance* (St. Martin's, 1989),

> The ugly fact is that America can no longer afford to provide adequate medical care for all its citizens. This means that some people will miss

> out on care they need. Some will die. And the problem will grow worse, not better, as the giant Baby Boom generation reaches its years of declining health.

American manufacturers, especially large manufacturing companies used to providing their workers with good, solid health care benefits, are becoming hard-pressed to keep up with costs. The cost per American for medical care in 1988 was $2,135, for a total national expense of more than $540 billion—about twice the cost of national defense. The average increase in health insurance in 1988 was 20 to 40 percent, with major cost drivers including high-tech medicine itself, malpractice insurance, and lawsuits. Bone marrow transplants, liver transplants, and heart transplants are all available now—at between $100,000 and $200,000 each.

Pharmaceuticals are also much improved, and much more expensive. Consider AZT for battling the AIDS virus, which costs roughly $6,500 per year, and the anticlotting drug TPA, which runs about $2,200 per dose. Health care benefits ultimately have a very tangible impact on our ability to competitively price our products. According to Richard Brennan in his book *Levitating Trains & Kamikaze Genes* (Wiley, 1990),

> The price tag of a typical new American car now includes about $360 in employee health care benefits. In contrast, the health care factor in the markup for an average Japanese car is only about $100.

Social Influences

Social influences, such as the diminishing impact of collective bargaining and the shifting demographics and values of our modern workforce, have come to wield tremendous power in the molding of American industry. Intertwined in the workings of the society that provides our workforce and (in most cases) our primary markets are factors such as the decline of the nuclear family, poverty, and inadequate education.

An Uneasy Fit: Today's Workers—Tomorrow's Jobs

A good place to begin when overviewing this topic from a manufacturer's perspective is with a simple demographic projection. Between now and the year 2000, two-thirds of all new hires in the United States will be female, and about 85 percent of new hires will be either female or minorities. The social factors that brought us to this point were largely evident for many years, yet we find ourselves largely unprepared to deal with them.

There is no question about the surging divorce rate and the resulting percentage of single-parent households in the United States. This situation has left a tremendous need for income in these families and has altered significantly the needs of our current and future workforce. Poverty rates for children in female-headed families was over 54 percent in 1986, while the rate for all other families was about 11 percent.

The numbers also tell a more ethnically oriented story. The poverty rate for white children in 1986 was about 18 percent, while black children suffered at a rate of about 46 percent and Hispanics at about 41 percent. Of course, the education of these children, some of whom inevitably join our nation's workforce, also suffers. William Johnston and Arnold Packer wrote in the Hudson Institute's *Workforce 2000* (1991),

> The workers who will join the work force between now and the year 2000 are not well-matched to the jobs that the economy is creating. A gap is emerging between the relatively low education and skills of new workers (many of whom are disadvantaged) and the advancing skill requirements of the new economy.

In 1982, American eighth graders scored in the bottom half of eleven nations competing in standardized math tests. A study by Chester Finn and Diane Ravage ("What Do Our 17-Year-Olds Know?") in 1987 showed that less than a third of our high school seniors can place the Civil War within fifty years. And by today's standards, that almost represents success. According to Cetron and Davies (1989),

> Nearly 1 million high school students drop out each year—about 30 percent of the total, on average, throughout the United States—and in some school districts the dropout rate exceeds 50 percent. Perhaps 700,000 more students finish out their twelve years hardly able to read their own diplomas. At that rate, by 2000, the literacy rate in America may be as low as 30 percent.

This situation places the responsibility increasingly on employers.

Beyond educational assistance, since our workforce is so rapidly shifting toward women and single parents, employers (including manufacturers) are also more frequently being asked to assist in some way with day care. Again, this can be an expensive undertaking. Average day care costs for most preschool programs, for instance, run from $50 to $150 per child per week. Nonetheless, by 1989, over 2,000 of the nation's 6 million employers provided some kind of day care arrangement as part of their employee benefits programs. While there is already evidence to suggest that such programs are related to lower employee absenteeism and higher worker productivity, their adoption is painfully slow.

More pervasive demographic trends will define the manufacturing workforce as well. The general population and the workforce are growing at their slowest rate since the 1930s. Average worker age is rising, and the pool of young workers entering the workforce is shrinking. While the overall impact of immigration (both legal and illegal) is uncertain, all of these factors spell real changes ahead.

Looking at the effects of these social factors on American manufacturing organizations, we see companies unprepared in most cases to deal with the training issues that must be addressed for this increasingly diverse and poorly educated workforce. They are unable, in many cases, to sustain the high cost of bringing this group up to the required level of education and skill themselves. In addition, most have not considered the long-term need for broader, more comprehensive employee support structures, including elements such as on-site day care.

Changes in Values

Another social factor impacting American business, and manufacturers specifically, is the value set shift in younger Americans. This is reflected in several ways: a desire for more autonomy in decisions pertaining to daily work direction, more interest on the part of employees in free time and sabbaticals than in promotions and raises, and so on. At the same time, this group hasn't reduced its spending levels. In fact, we still have an escalating rate of consumer debt and a declining rate of personal savings. Personal savings (expressed as a percentage of disposable income) has declined steadily over the last decade. It is the lowest, in fact, of any major industrial country. This means, of course, that fewer funds are available for investment in capital expansion and R&D.

The "Green" Consumer

Social awareness as it relates to environmentalism is still another major influence on current manufacturing operations and is likely to grow in strength in the future. Manufacturers here and abroad have been expelling toxic chemical wastes into the atmosphere, into the water, and onto the ground for about 200 years. The chemicals include mercury, sulfuric acid, chlorine, and even cyanide.

Recently, due largely to increased pressure from environmental groups, the federal government has "encouraged" manufacturers to police their discharged materials, and has even provided incentives to look into recycling them. One of the most promising of such techniques reported recently is chlorinolysis, a process used by Dow Chemical in California to reuse waste streams from agricultural chemical plants.

There is also a real image-boosting value available to industrial companies that publicly embrace environmentalism. Companies including Du Pont, McDonald's, 3M, Procter & Gamble, and Pacific Gas & Electric are all spending vast sums of money to shift toward an environmentalist posture in the eyes of the public. There is little doubt that government regulation will continue to tighten as social pressure increases. It is a message manufacturers are finding increasingly difficult to ignore.

Competitor Influences

Competitor influences, such as competitor involvement with consortia and industrywide effects from disasters such as Bhopal and the Tylenol poisonings, can be sudden and transformational. The enormous expenditures made for safer packaging after the poisoning incident involving Tylenol demonstrate that even companies doing their best to deliver safe, high-quality products sometimes face unfair disadvantages relative to their competitors.

Turning Competitors Into Partners Through Strategic Alliances

Consortia and other forms of joint business ventures have historically been retarded in their growth by suspicion on the part of the American public as well as the federal government. Yet this means of turning competitors into partners has been proved to be a powerful way to achieve technological breakthroughs and to spread the benefits of large contracts through "teaming" arrangements.

Two fairly compatible fields for developing these relationships are the areas of training and education and research. Universities and departments of government, often competitors, can be teamed very successfully in these areas.

A fairly strong recent trend, particularly among smaller entrepreneurial firms seeking to compete among giants, is the "strategic alliance." A recent study of more than 1,100 biotech companies, for example, reported that over two-thirds of respondents to the survey were involved in some form of strategic alliance, primarily to improve their marketing capabilities. Alliances among competitors often strongly affect several areas, depending on the nature of the alliance.

Broadly defined, strategic alliances have been developed in these five categories:

1. *Equity-based alliances,* typically involving minority ownership or directorship in one company by another, in order to codevelop products or processes
2. *Manufacturing alliances,* usually designed to provide manufacturing capability to a company whose needs exceed its capacity, or to one that has no manufacturing capability at all
3. *R&D alliances,* generally involving an arrangement to provide financing for R&D in return for a share of later profits
4. *Marketing alliances,* usually designed to provide marketing and distribution capabilities to a company that has none of its own
5. *Training and education alliances,* designed to promote common technology development and industry expertise

Accepting the Competitors' Challenge

Productivity differences of a significant magnitude have an important effect on the cost and pricing structure of manufacturing competitors' products and ultimately on their overall viability in the marketplace. Witness the advent of Japanese companies in American markets over the last decade. A look at the *Statistical Abstract of the United States,* published by the U.S. Department of Commerce, shows that the output per hour of American manufacturers between 1970 and 1986 rose by more than 57 percent, while the same output for Japanese manufacturers shows a 159 percent increase, with Japan first surpassing the United States in output per hour in 1980.

Technology improvement on the part of competitors has also proved to be a potent force. The new technology may be in either processes or the product itself. (Consider the plight of typewriter manufacturers under the onslaught of word processing, and the evolution from phonograph records through eight-track and cassette tapes, and most recently to compact discs.) Manufacturers in these situations have been and are continuing to be faced with reacting to them by either attempting to incorporate the new technology themselves, developing competing technologies, developing complementary products and processes, or resigning themselves to lost market share.

This leads us to the recently prominent concern of intellectual property protection. New technology is becoming more difficult to protect as broader global capability exists to "reverse engineer" products, and information flow is not as easy to restrict. Anyone who doubts the difficulty of protecting even the most sensitive information in today's environment of electronic data interchange is encouraged to read *The Cuckoo's Egg* by Clifford Stoll. Especially with the advent of computer aided design (CAD) and computer aided engineering (CAE), larger and larger amounts of confidential and even classified data are vulnerable to anyone with adequate systems knowledge, patience, and a personal computer with a modem.

The darkest side of having access to information about one's competitors is the temptation to use it to hurt them in unethical and even immoral ways. Today, this kind of behavior is seen in the production and deployment of computer software "viruses." Imagine the potential effects of a sociopathic mind gaining access to software that controls the chemical formulation of nonprescription drugs in a competitor's factory. Today, random acts of one deranged mind tainting bottles of pain reliever can bring a company to its knees (at least temporarily). In the future, technology will offer more opportunities for more complex and widespread disasters—particularly for the most technically skilled. Heretofore, random disasters of the magnitude of the Tylenol poisonings and the Bhopal debacle will be "engineerable" by vicious and unethical competitors.

Enabler Influences

Enabler influences are the most fundamental of environmental influences. They are powerful discoveries, innovations, or theories that enable mankind to make significant developmental leaps, often across a broad band of disciplines. (The invention of the printing press and the discovery of radio waves are two examples.)

Making a Quantum Leap

I believe that the most potent enabler in recent years has been quantum physics. Through landmark discoveries in this field, nearly every major technology has benefited. Quantum physics has enabled us to produce new manufacturing materials, develop enhanced purification and chemical processing methods, and make enormous strides in computing and communications technologies.

There is little doubt that quantum physics is yielding impressive results in a broad spectrum of fields. Texas Instruments, for example, has built a "quantum effect" transistor that harnesses subatomic forces in order to push the current limits of "smallness" in reliable transistor-based computation. In fact, it was quantum theory that first revealed that materials such as silicon and germanium are versatile enough to act as either conductors or insulators, thereby giving rise to devices such as silicon and gallium arsenide semiconductor chips. It is also the basis for our understanding and recent progress in the field of superconductors, as well as some of the most startling recent advances in biotechnology.

It's impossible to say exactly how far this particular "enabler" will eventually take us, but it will certainly change our perception of reality.

Summary

Recognizing the tremendous impact and volatility of environmental influences, we will be increasingly required to project potential changes in these areas, and build as much flexibility as possible into our organizations. In this manner, we can be in the best possible position to anticipate and respond to them as they occur. In some cases, we may choose to develop new products and processes in anticipation of an environmental change. In other cases, we will be able to build environmental change immunity into existing products. Still other cases will no doubt warrant a change in the markets we choose to serve and the product lines we choose to deal in.

These insights will have important ramifications on all of a company's business processes, with particularly heavy influence on the business administration and distribution processes. The need for a structured change management process and change support procedures will become paramount to success. The ways in which manufacturing companies are likely to anticipate change and incorporate flexibility to accommodate it in the decade ahead is a subject of discussion in Chapter 12.

Chapter 4

Dominant Manpower Issues: The End of the Direct Labor Era

Workforce demographics have been a very interesting study over the last three decades. They reflect significant changes in our society as a whole, and in the needs of our workers (see Exhibit 4-1).

Male and Female . . . Female and Male

With the exception of 1978, the percentage of the American labor force comprised of adult men has steadily declined during this period. In fact, 1983 was the first time in recorded history that white men made up less than half of our workforce. Conversely, the percentage of adult women in the American labor force grew from about 33 percent in 1955 to 55 percent in 1985.

White women comprised the lion's share of this growth, especially women between ages 25 and 34. Also between 1970 and today, the number of women in the labor force with children under age 18 increased much faster than the total female participation. By 1986, the labor force participation rate for this group of mothers was about 62 percent. (However, about half of all working women do not have full-time, year-round jobs.)

The importance of the shift from men to women in workforce dominance is still a contested issue, but women are frequently described by business publications in terms such as "the linchpin" in the transition from manufacturing to a service-based economy in the United States. That last point is interesting for at least two reasons:

1. The economy is very definitely shifting away from manufacturing toward services, which we'll discuss in more detail shortly.

2. Even within manufacturing, and especially in support functions surrounding manufacturing, women have been gravitating toward specific areas. This is at least coincidental, and may well be causal of some trends we have seen recently. For example, it may show an old bias against women in "direct labor" roles. However, it reflects a situation where women are increasingly occupying positions where product-shaping and process-shaping decisions are made. This will almost certainly help to "even out" the built-in biases in downstream operations. If it *is* causal, and the trend of participation by women escalates at projected rates, we may foresee some interesting developments.

Exhibit 4-1. Dominant manpower issues affecting American manufacturing.

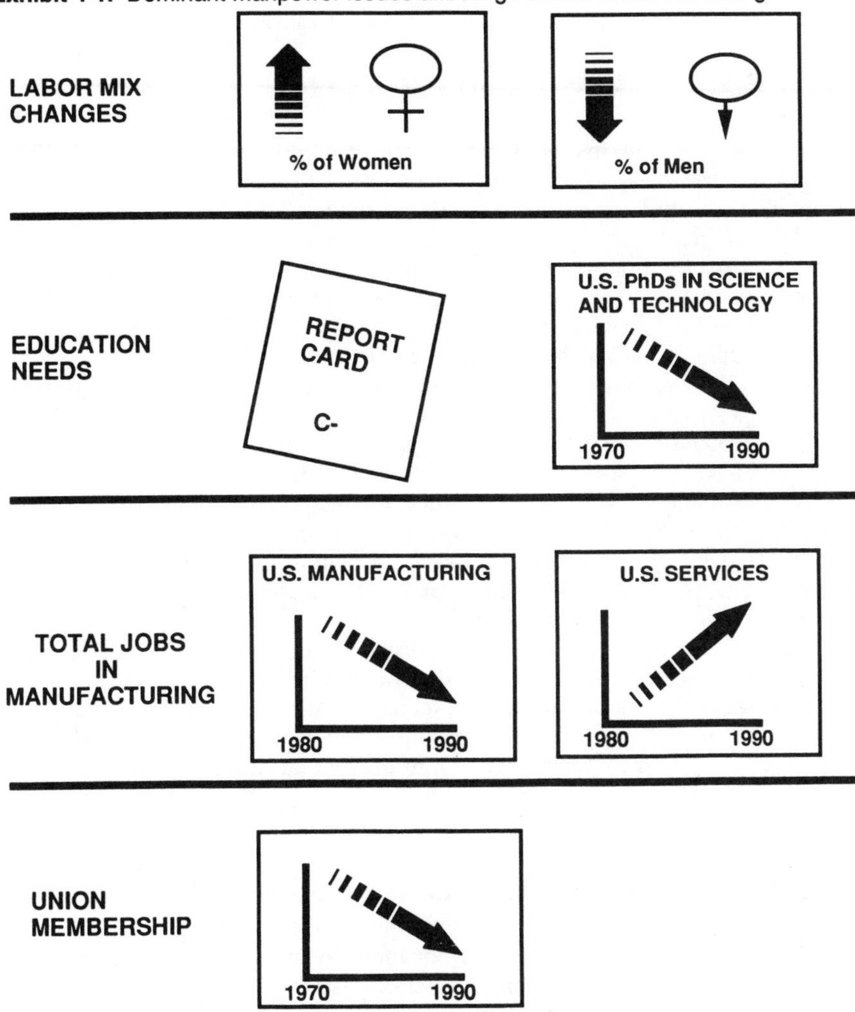

Some specific manufacturing-related positions women are concentrating in include:

- *Electrical and electronics engineers,* where the participation rate has more than doubled (to about 10 percent) over the last decade.
- *Industrial engineers,* where the participation rate grew from 9 to 16 percent over the same period.
- *Computer systems analysts* (from 20 percent to 30 percent) and computer programmers (from 28 percent to 40 percent).
- *Accountants and auditors,* where participation rates increased from 34 percent

to 45 percent. (Of related interest, in 1975 about 13 percent of all M.B.A. graduates were female. By 1985, the figure was 33 percent.)

Black Participation: Opportunity Lost

The demographics of the black population also influences the manufacturing workforce. First of all, black adult males' participation rate in the workforce generally runs significantly lower than that of white adult males.

For example, according to the *1988 Statistical Abstract of the United States*, there were about twenty times as many white male managerial/professional employees in the United States as there were black males in that category. Even when only adult males with four-year degrees were compared, the white male figure was larger than that of black males by about twenty times.

In addition, black male workforce participation appears to have dwindled in recent years and is projected to continue to diminish.

Black women have fared better than their male counterparts, with roughly the same level of female participation for black women and white women in the labor force. However, even with similar participation rates, the total income of black women is still less than that of their white counterparts.

These elements have several important implications for manufacturers. For example, America's black population represents a large percentage of the domestic market for such products. Continued low levels of black participation in the workforce would mean that we lose the ideals, views, and values of this group in the design, development, and marketing of products. That translates into lost market share and a narrow view of market opportunities. As we move through the 1990s, we are faced with serious questions about how to draw qualified people from black and other minority groups into the manufacturing world with greater frequency and to effectively incorporate their attributes into our businesses—to the benefit of *all* of us.

Industry and Education: Tentative Beginnings

Education and training have become hot topics over the last two decades, and education specifically is generating heated discussions between and among both industry professionals and academics.

The good news is that the participation of industry in educational policy setting and educational innovation has grown to an unprecedented level recently, as evidenced by the publication of two fascinating books on the subject, written and cowritten by industry figures. One of these is *Winning the Brain Race* (ICS Press, 1988) by David Kearns, the CEO of Xerox Corporation, and Denis P. Doyle, a research fellow at the Hudson Institute. Kearns lays out a straightforward plan of "free market schooling," restructuring the public school system itself, offering enhancement incentives for teacher professionalism and performance, raising academic standards, inculcating democratic values, and improving the role of the federal government in education.

The other book, while less popularly received, is no less important. *The Seven Ability Plan* is written by a former colleague, Arnold Skromme, who is now retired from the engineering department at John Deere. This ground-breaking book demonstrates the need to revamp the curricula and criteria of current education. It suggests, for example, that students be graded on not only academics but also on creativity, dexterity, empathy, judgment, motivation, and personality. Although the book is centered around a "7 Ability Plan," it clearly emphasizes creativity.

There are several interesting things about this recent trend of interest, involvement, and deep commitment on the part of industry (and for our purposes, manufacturing specifically) to education. First of all, it is an interesting coincidence that Xerox, whose CEO is so committed to educational excellence, was the recent winner of the Malcolm Baldrige National Quality Award—one of the most coveted and prestigious awards a manufacturer can receive. Second, although it is possible to disagree with the specific prescriptions of the authors, it is nonetheless striking to see manufacturing professionals so concerned about what has not historically been seen as a manufacturing issue. This is a very positive sign of our increasing ability to recognize the potential impact of broad national and international changes and to proactively address them for our long-range benefit. We are definitely growing!

The bad news is that the current state of American education and training is deteriorating, which does not bode well for American manufacturers. Competitors like Japan are providing a much more intense educational environment, and hence producing higher test scores.

In addition, the percentage of Japanese students who complete high school is somewhere around 94 percent, while the U.S. figure is around 71 percent and has fallen for more than ten years. The Japanese school year is 240 days rather than 180, and Japanese students graduating from high school are educated at a level roughly equivalent to three years beyond the U.S. average.

Because Japan's culture is much more highly structured, their more intense approach is far better suited to Japan's environment than to America's. It seems that Kearns's approach of free market schooling (*Winning the Brain Race*) would be more successful in the United States.

However we design our response, America must respond quickly and vigorously to the challenge of global competition for the best-educated and best-skilled workforce. Throughout previous decades, when the bulk of manufacturing jobs were "touch labor" jobs, very little was required beyond fundamental reading and math skills. As these jobs disappear, they are being replaced by jobs that require computer skills, higher math skills, abstract reasoning ability, and interpersonal skills. Even the remaining "touch labor" employees are now required to do statistical process control, operate computer terminals and automated identification equipment, and work in cooperative "team" settings with other employees. These are not skills traditionally imparted in America's educational system, but they will become as fundamental as reading was twenty years ago.

Today, American manufacturers are attempting to compete in global markets through the application of artificial intelligence, sophisticated production planning and control systems, complex modeling and simulation techniques, and light-speed telecommunications. At the same time, these efforts are critically undermined by our own inability to provide an adequately educated and trained workforce. According to a 1985 report by the National Center for Education and Statistics on seventeen-year-old American males, almost 40 percent could not draw inferences from written material, and over 50 percent could not comprehend such basic scientific concepts as gravity.

The Shifting Sands of Manufacturing Employment

The job mix in American industry has changed dramatically over the last two decades, mostly from manufacturing to service jobs. About 75 percent of the new jobs created in the 1980s were among the lowest-paying job categories (retail and health services).

In 1986, according to the U.S. Bureau of Labor Statistics, here was the breakdown of manufacturing employment:

TOTAL MANUFACTURING	18,994,000
Durable Goods	11,244,000
Lumber and wood products	711,000
Furniture and fixtures	497,000
Stone, clay, and glass products	586,000
Primary metal industries	753,000
Fabricated metal products	1,433,000
Machinery (except electrical)	2,059,000
Electrical and electronic equipment	2,124,000
Transportation equipment	2,016,000
Instruments and related products	707,000
Miscellaneous manufacturing	362,000
Nondurable Goods	7,750,000
Food and kindred products	1,617,000
Tobacco	59,000
Textile mill products	706,000
Apparel and other textile products	1,105,000
Paper and allied products	675,000
Printing and publishing	1,458,000
Chemicals and allied products	1,023,000
Petroleum and coal products	169,000
Rubber and miscellaneous plastics	789,000
Leather and leather products	152,000

The overall trend of employment in manufacturing over the last two decades overshadows specific industry figures. For example, the total manufacturing number declined from 20,286,000 in 1980 to 18,994,000 in 1986. Durable goods went from 12,188,000 to 11,244,000 in this same period, and nondurable goods fell from 8,098,000 to 7,750,000.

The occupational mix is not all that has changed in this area. As a rule, manufacturing wages have run about 30 percent higher than wages in the service sector of the economy over the last decade. However, because so many service jobs involve shorter work weeks than manufacturing jobs, the real difference in income between the two job categories averages about 50 percent. (The average weekly wage for manufacturing in 1986 was about $400, while the average for retail—the largest nonmanufacturing segment—was about $175.)

Within manufacturing, the typical job distribution of people in mainstream companies involved in both fabrication and assembly operations looks something like this:

Primary Function	*Approximate Percentage*
Business administration	22%
Obtaining business	5
Defining products and processes	24
Production	32
Distribution	7
Support	10

As stated in our discussion of environmental factors, the last two decades have seen manufacturing, design, and administrative functions become more complex as markets have become increasingly competitive. Hence, emphasis has steadily shifted away from the traditional production and distribution functions to the business administration and design areas. In subsequent chapters, we will examine how this trend is likely to evolve over the next several years.

Warning: Service Economies Are Hazardous to Your Economic Health!

The economic shift from manufacturing to services is dangerous for several reasons. The most important and most distinctive aspect of this trend is that U.S. manufacturers are, for the first time on a large scale, transporting our most valuable commodity—our technology—overseas. Not only production technology is involved here but also more critical elements such as design, materials engineering, systems, and management skills.

In addition, national defense, even in these times of "peace breaking out everywhere," is extremely important. It is important socially, economically, and politically. Exporting our most valuable technical skills and production technology to potential adversaries isn't simply unintelligent; it's naive. History holds some clear lessons for us regarding the speed and abruptness with which our international friends can become wartime enemies.

Unions: Origins and Outlook

Unions have played an important part in the development of American manufacturing between the 1930s and today. They grew dramatically between the mid-1930s and the mid-1940s, enjoyed great stability in terms of membership until the late 1960s, and began to seriously decline in the late 1970s. Peak membership years for most unions were between 1969 and 1976. In the first half of the 1980s, union membership fell by more than 2.5 million, decreasing from 23 percent to 19 percent in that period.

Specific unions heavily involved in manufacturing include the following:

- *The Teamsters,* whose peak membership year was 1974, with 1,946,000 members
- *The United Auto Workers,* whose peak membership year was 1969, with 1,426,000 members
- *The United Steel Workers,* whose peak membership year was 1975, with 1,071,000 members
- *The International Ladies' Garment Workers,* whose peak membership year was 1969, with 387,000 members
- *The International Brotherhood of Electrical Workers,* whose peak membership year was 1975, with 856,000 members
- *The International Association of Machinists,* whose peak membership year was 1975, with 780,000 members

Virtually all of the major unions have lost membership over the last ten years. As of 1987, for example, the United Auto Workers were listed at 998,000 members (a 30 percent decline), the United Steel Workers were showing 494,000 members (a 54 percent loss), and the International Brotherhood of Electrical Workers had 765,000 members listed, for an 11 percent drop.

Unions were also forced to accept a number of serious concessions throughout the 1980s, with the most prominent and consistent defeats occurring in the area of wages. In 1980, less than one percent of union members were accepting wage cuts. By 1983, about 37 percent had accepted such losses in order to maintain their jobs.

The role of unions in modern American manufacturing is increasingly brought into question. The problem management faces with unions today is more often related to job classifications than it is to wages. As our competition is increasingly global in nature, we face increasingly flexible and "hungry" competitors. We often constrain ourselves by doing things like requiring different job classifications to move parts between operations because they cross departmental boundaries, even though the move is sometimes only a few steps. These practices, while saving a few jobs in the near term, make us less competitive and end up costing thousands of jobs lost to less constrained and therefore more efficient competitors.

Even so, it is important to keep the situation in perspective. In most cases, the real cost attributable to direct labor in American manufacturing is somewhere around 10 to 15 percent. Yet we beat this element of our cost mercilessly, because it is visible and easy to measure. Material costs are easy to measure, but more difficult to affect. And the most fertile area of all, overhead, is often largely ignored. We don't know how to measure its efficiency, and, frankly, who would want to? Very few of us would like to undergo the productivity measures we often inflict on our direct labor force.

Chapter 5

Dominant Methods: Doing It by the Numbers

A number of far-reaching factors have influenced the evolution of methods used in manufacturing as a whole, ranging from management theories to materials technologies (see Exhibit 5-1). The rudiments of the currently dominant methods going through the first half of the 1990s are reflected in Exhibit 5-2.

The Methods Used in Business Administration

Among mainstream American manufacturers the primary functions of business administration include strategic planning, budgeting and business planning, accounting and cost management, human relations and employee benefits, management information systems and telecommunications, and change management.

Strategic Planning: A Constant Star on the Manufacturing Horizon

Strategic planning, budgeting, and business planning are currently done for a five-year horizon, with yearly iterations. They normally include a strategic industry analysis, which differs for most companies from the plans that preceded it in at least one respect: Strategic industry analyses today are far more focused on foreign competitors and their effect on the market. (Unfortunately, most of the energy expended in this area is devoted to reactionary measures such as lobbying for trade restrictions rather than proactive efforts related to productivity improvement.) The strategic industry analysis typically pays only cursory attention to far more vital environmental issues such as technological developments, political scenarios, and demographics.

Strategic business analyses and budgets are the other aspects of current strategic planning efforts. Generally, the analyses center around a method that views all data in terms of strengths, weaknesses, opportunities, or threats ("SWOT"). Again, recent planning efforts have at least one major change in focus from preceding efforts: There is a far sharper interest in protecting the company from hostile takeovers and heightened interest in mergers, acquisitions, and divestitures. The interest in mergers and acquisitions would not in itself be so troubling if these corporate marriages made good business sense for both partners. As noted earlier, often they do not.

For the most part, the assumptions and underlying goals of strategic business plans have remained the same for many years: high market share, high capacity utilization in terms of both facilities/equipment and labor, and of course maximum

(Text continues on page 36.)

Exhibit 5-1. Major influences on American manufacturing methods.

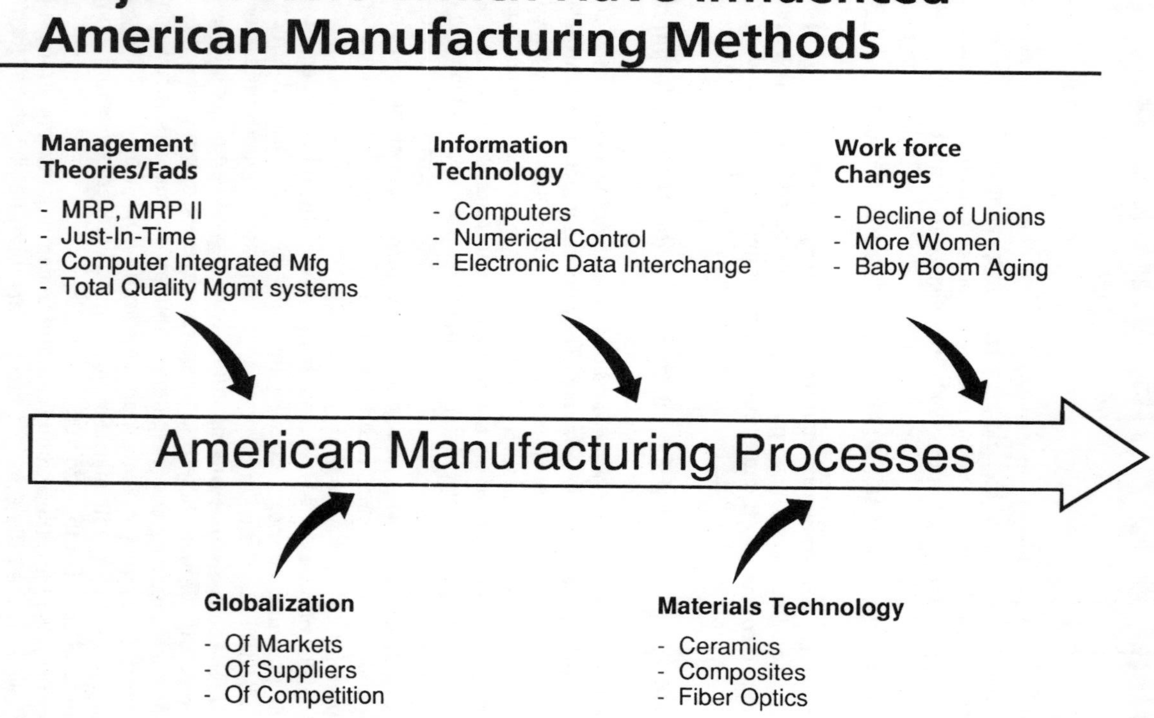

Exhibit 5-2. Dominant U.S. manufacturing methods.

Dominant Methods

U.S. Manufacturing Methods

Business Administration	Obtaining Business	Defining Products and Processes	Production	Distribution	Support
Strategic Planning Budgeting and Business Planning Accounting and Cost Management Human Relations and Employee Benefits Management Information Systems Change Management	Pricing Practice Changes Product Alterations to Meet Diverse / Rapid Market Changes Cycle - To - Market Streamlining Widespread Application of Automation in Promotion Activities	CAD/CAM Quality Function Deployment Value Analysis / Value Engineering Concurrent Engineering Group Technology Design for Manufacture Automated Process Planning	Production Management: - Production Planning & Control: - MRP, MRPII - Just In Time - Computer Integrated Manufacturing Quality Assurance Shop Floor Supervision Conversion: - Metal Forming - Metal working and Forging - Metal Removal - Casting - Molding - Treating - Plastics Processing - Assembly	Internal Stocking and Controls: - Forecasting - Order Management Logistics: - Route Planning - Field Management - Inventory Management and Redistribution Distribution Center Management: - Electronic Data Interchange - Inventory Management - Automated Storage and Retrieval Systems	Repair Part Delivery: - 3rd Parties - Customer Order Entry - Electronic Data Interchange Field Service: - 3rd Parties - Concurrent Engineering - Assembly Line Training Returns: - 3rd Parties - Exchanges - Same-day Service Warranty Management: - Maintenance Agreements - Preventive Maintenance Technical Publications: - Desktop Publishing - Computer Aided Design Field Performance Data Management: - Electronic Data Interchange - Computer Analysis

The methods used in business administration.

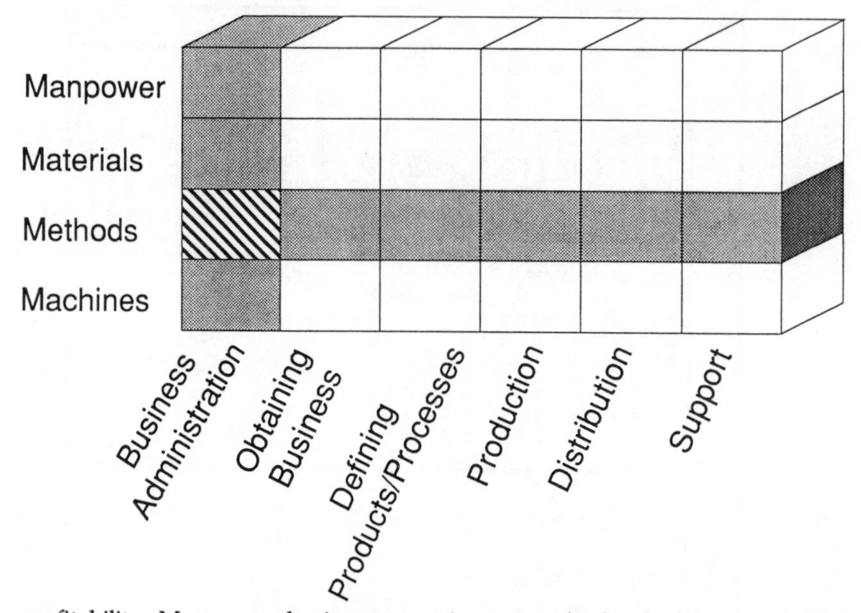

profitability. More recently, investment intensity (the level of fixed and working capital required to produce a dollar in sales) and direct cost per sales unit figures have been appearing with more regularity as well.

Accounting for Global Change

Accounting and cost management have been areas of special management attention in the last three to five years. Manufacturers in particular have come to realize that global competition will require far better cost management than has previously been typical. Consider the cover story of *Fortune*'s April 9, 1990, issue, entitled "Cost Cutting: How to Do It Right":

> After nearly a decade of slashing overhead and slicing jobs, corporate America has entered the Nineties ready to slash and slice some more. The list of Companies that announced work force reductions in the past six months alone would fill this page. A few of them: Apple Computer, AT&T, Bank of New England, Boeing, Businessland, Caterpillar, Chrysler, Eastman Kodak, Gillette, Grumman, Honeywell, Hughes Aircraft, Levi Strauss, Merrill Lynch, McGraw-Hill, US West Communications.

Today, cost accounting really does seem to be evolving into cost management. It involves a much more comprehensive view of cost, including previously unexplored territory such as "cost of quality" (COQ). It involves continuously identifying and eliminating waste and focusing on what my friend Mike Ostrenga at Ernst & Young refers to as "cost drivers." In many cases, manufacturing management is waking up to the fact that direct labor is a minor part of their cost, and beating it to death while ignoring overhead and capital expenditure simply isn't working.

Financial reporting is of course still driven by generally accepted accounting principles (GAAP) and remains largely the same in format. Things that *have* changed in this arena include the following:

- The reporting format (moving from reporting changes in financial position to a cash flow statement), per Financial Accounting Standards Bulletin (FASB) 95
- Reporting about pension funds (creating a pension credit and changing the valuation methods and disclosure methods related to pension funds), per FASB 87 and 88
- Income tax reporting (going from a liability method to a deferral method), per FASB 96 (not yet in effect as of this writing)
- A requirement to combine all major subsidiaries for reporting purposes, per FASB 94

These changes primarily add more layers of complexity to the financial reporting aspect of business administration. They have caused particular confusion and required the most professional auditing assistance in the tax and pension areas.

Employee Benefits à la Carte

Human relations and employee benefits functions have become an increasingly critical aspect of business administration since the 1960s and promise to continue to grow in importance over the next ten to fifteen years. Many of these changes are demographic in origin, as stated in Chapter 4. Because of the recent infusion of single parents and dual income families, there is more mobility in today's workforce and less longevity.

An interesting cover story appeared in an August 1990 issue of *Fortune* magazine that describes in detail some of the radical differences between the entry-level workers of the 1990s (coined yiffies) and the preceding generation of yuppies. The changes appear to stem from a fundamental change of values. Perhaps as never before, new workers appear less materialistic, less willing to make big sacrifices (such as moving to a distant city) for their careers, and less interested in accepting a more stressful job in return for higher pay. They are big believers in themselves and in the theory that "what they know" will make them valuable. They search for diversity of experience rather than a more traditional and focused climb up the corporate ladder.

Unfortunately, as noted in Chapter 3, "what they know" is increasingly meager. Already, major manufacturing companies are reporting severe shortfalls of highly educated categories of workers such as scientists and engineers. One study quoted by *Aviation Week & Space Technology* (January 1, 1990) reported that two-thirds of the companies surveyed had skills shortages in areas such as advanced materials, electro-optics, and artificial intelligence.

Similarly, manufacturers' benefits programs must become as diverse and responsive as the employees they serve. Educational benefits, day care, leaves of absence, and job sharing are becoming increasingly common. Human relations professionals are finding these trends difficult to manage in terms of their diversity and have most recently responded with programs called "cafeteria style" plans, in which employees can literally choose the benefits most suited to their individual needs, spending "credits" allocated by the company to each employee. Some employees choose to get more vacation time, while others opt for more comprehensive medical or life insurance coverage, for example.

Missed Opportunities in Management Information Systems

Management information systems (MIS) and telecommunications have, of course, washed across manufacturing organizations in relentless waves over the last thirty years. It has been a love/hate relationship in most cases, and even now these systems are regarded with skepticism in many organizations. Part of the reason is that management placed unrealistic expectations on their systems early on and were disappointed. It was a hard lesson to learn (or relearn) in this light-speed world that "garbage in" really does mean "garbage out." Errors input to manufacturing systems traveled with unparalleled speed to every corner of the operation, sometimes with disastrous results.

As a production control scheduler at John Deere in the 1970s, I remember discovering one day to everyone's great surprise that a simple data entry error had virtually wiped out all of the manufacturing orders in the factory. The error? A transposed part number that told the computer we had dozens of complete combines in a single four-foot-square container in the high-rise storage facility. We have learned a lot as manufacturers about the value of good systems edits since those days.

Another real problem plaguing our information systems today is a general lack of standardization. Many industries have attempted to adopt rudimentary standards for electronic data interchange, such as the ANSI (American National Standards Institute) X.12 and AIAG (Automotive Industry Action Group) standards. However, the standards are neither comprehensive nor universally accepted and therefore not standard at all.

An additional and virtually universal challenge for management is systems integration. Even the best of the current MRP II software must ultimately be bridged to modern design (i.e., CAD/CAM) systems, machine control (i.e., NC) systems, and often to supplier and customer systems. The last twenty years have seen an explosion in the growth of electronic data interchange (EDI). In part because of the declining costs of EDI hardware and software, and spurred on by the drive toward just-in-time (JIT) procurement and manufacturing, data sharing between suppliers and customers has reached unprecedented levels. Again, one of the hindrances to progress in this area is the current lack of data standards.

Types of Data Required

Despite the foregoing obstacles to effective MIS, electronic data transmission is quickly becoming the rule rather than the exception. Among the kinds of data being moved commonly over telephone lines and bounced off satellites today are the following:

- Performance data from field operations
- Sales data from distribution centers and points of sale
- Design data from and to suppliers
- Production forecast data to suppliers
- Schedule revisions to suppliers
- Service and support requirements from field support operations
- Purchased parts requirements to suppliers
- Shipping information from suppliers and carriers
- Marketing and competitor intelligence from commercial on-line data bases
- Credit information about potential customers and suppliers from commercial on-line data bases
- Electronic funds transfer (EFT) between:
  —Customer and manufacturer
  —Manufacturer and supplier

The Need for Data Protection

Of course, all this data interchange is posing real problems in terms of data security. A rough estimate by the accounting firm of Ernst & Young in 1988 put the cost of computer crime at "between $3 billion and $5 billion a year." (February/March, *Technology Review*).

Beyond deliberate theft and sabotage by hackers, there is also an increasing risk of plain old systems breakdowns. *Bugs* in systems, long-dormant *viruses*, *glitches*, and a host of other terms in the systems vernacular have been adopted over the last two decades to describe these phenomena. As manufacturers increase their dependence on EDI, they increase their vulnerability. When AT&T's long-distance telephone network broke down for nine hours in 1990, only 50 percent of the 148 million long-distance and 800-number calls made got through. How many of those calls were manufacturers desperately needing to exchange data within their own organizations, or with customers and suppliers?

With all of these challenges and opportunities manifesting themselves, senior management in some manufacturing organizations over the last few years have created a new corporate officer position dubbed chief information officer (CIO). *Business Week* first reported this development in an article entitled "Management's Newest Star—Meet the Chief Information Officer" in its October 13, 1986, issue.

Recent press has decried that movement, but the jury is still out. There is little doubt that the complexity and importance of management information systems have grown at an incredible rate since the 1960s. The ability to quickly turn raw data into useful information and effectively communicate it has become the hallmark of successful managers in every area of business. Manufacturing is no exception.

Change Management as a Survival Tactic

The most crucial function of current and future managers is change management. The methods used by manufacturers to incorporate positive change and deal with negative effects of change will determine which of them survive. Robert Gilbreath, in his book *Forward Thinking* (McGraw-Hill, 1987), states that there are four levels of reaction to change: ignorance, recognition, accommodation, and control. He also does a good job of describing several ways to accommodate and control change, such as designing "timeless" products that either can be adapted to the environment as it changes, or making them so enduring that they resist the changes that occur around them.

Manufacturing organizations have not historically taken this "long view" of their products and processes. Management has looked more toward short-term profitability through exploiting whatever market niches present themselves, whether fad or not. Notable exceptions include Henry Ford and Steve Jobs—people who not only *perceived* and *controlled* coming change, but *created* it.

Far more commonly, change management has been centered around merely recognizing existing trends in demographics, technology, and markets and struggling to satisfy the perceived needs created by those trends with modifications of existing products and processes. Whereas American managers regard five years as the time frame for long-range planning, some of our more successful foreign competitors are working to plans that span decades.

The Methods Used in Obtaining Business

American manufacturing has made some real strides over the last two decades in the area of obtaining business (often described as marketing). The areas of major emphasis in this field are generally categorized under the headings of pricing, product, and promotion.

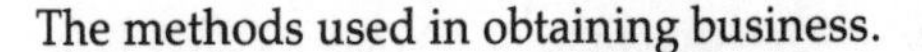
The methods used in obtaining business.

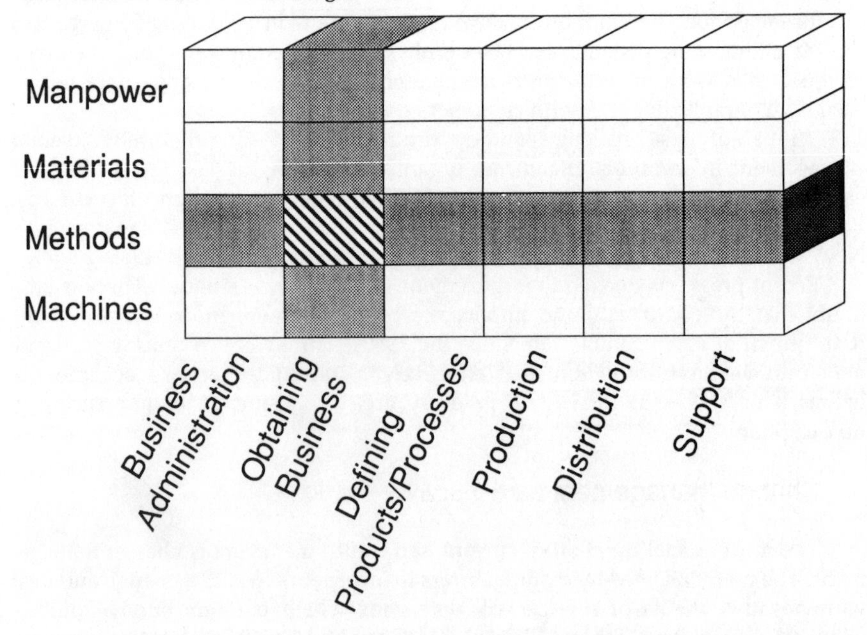

The Pricing Revolution

Pricing practices commonly involve discounting for especially prompt payment, quantity-related and other special discounts, and strategic market entry and domination by market niche or price range.

Two related, major developments have taken place pertaining to pricing practices (methods) of American producers in recent years. The first of these was the onslaught of Japanese (and later Korean) products. The pricing strategy employed by these competitors was truly ingenious, because it allowed for learning curves and minimum up-front investment while penetrating the most venerable aspect of the American market. The strategy was to enter our markets at the low end of the price spectrum, building sales volume and name recognition. American producers in most cases did not strongly contest this piece of their markets, choosing to focus instead on the higher end of the spectrum where margins were greater. The Japanese manufacturers then exploited their momentum and the experience gained by progressing down the learning curve to challenge and capture more lucrative market segments. Examples are plentiful in industries like copiers, machine tools, and semiconductors.

A case in point: While U.S. companies pioneered the manufacturing of high-end computer aided tomography (CAT) scanners in the early 1970s, Japanese businesses followed shortly thereafter with lower-priced, pared-down units that smaller hospitals could more easily afford. After their sales volume grew, the Japanese companies could afford to expand their sales and service efforts. They had sacrificed short-term profits for market share. Meanwhile, pressured U.S. companies boosted prices and cut back on marketing and sales-related expenditures. The benefits of this tactic were short-lived, and, by the late 1980s, the Japanese products had risen in capability and features to match the American offerings, while their prices *remained lower*.

American manufacturers did eventually wise up, and there are a number of stories to attest to that awakening. Now every market segment is hotly contested, and pricing strategies include competitive postures in even the low end of the price spectrum.

The second pricing method development is the more recent discovery by American producers (again largely as a result of foreign encroachment into what were wholly American markets) that consumers will pay more—a *lot* more—for higher-quality products. Defining quality as a combination of reliability, durability, maintainability, and user-friendliness, American consumers have become extremely quality-cognizant buyers. Therefore, higher price levels are being achieved for products with a high-quality image. According to an article entitled "How To Deal With Tougher Customers" in the December 3, 1990, issue of *Fortune,*

> Indeed, today's consumer is willing to spend more to get more. In Gallup's 1988 survey, shoppers said they would pay 72% above the $500 base price of a sofa for a better piece of furniture. In 1985, they would pay only 4% more. Gallup also discovered that by 1988 consumers who wanted quality would pay 67% more than the $300 base price for a TV, 42% above the $400 base price of a dishwasher, and 21% above the

$12,000 sticker price for a car. Three years earlier, quality was worth only an additional 6% for the TV, 3% for the dishwasher, 10% for the car.

Marketing's Increased Flexibility

The area of product alteration and adjustment as a part of the marketing process has undergone tremendous and varied developments in the last twenty years, and, consistent with all other areas of manufacturing, the change that is most rapid is also most recent. Among the initiatives of American producers in this field are the following:

- Formal programs of collaboration between marketing people and their customers to better understand customer needs.
- Focused programs on reducing order entry time to speed delivery to customers, using new technologies like electronic data interchange (EDI), to communicate order information immediately, directly from the customer site.
- Applications of automated information gathering for marketing purposes, such as building in computer chips that store operations records automatically in products so that sales and usage trends can be more quickly and accurately obtained.
- Frequent alteration of products based on early feedback about new product performance.
- Tailored packaging of products to satisfy rapidly diversifying market segments.
- Formal programs of rapid prototyping to satisfy changing markets, employing interdisciplinary teams.
- Marketing partnerships with suppliers to build supplier expertise into the product.
- Formal and regular customer review boards to recommend product changes and improvements.
- Adoption of company performance measures based on cycle-to-market of product changes and new products.
- Use of interdisciplinary teams to streamline the response time from development through distribution; one of the companies most heavily involved in these efforts recently is Procter & Gamble, which has developed teams composed of manufacturing, engineering, distribution, and purchasing personnel who report to product supply managers for this purpose.
- Satellite-based ordering of product to provide it on a just-in-time basis to retailers and customers; P&G again leads the field in this area, ordering disposable diapers just in time from the factory to be shipped to Wal-Mart stores.
- Environmental safety, as evidenced by the clear statement of environment-sensitive features on products (such as "Packaged in recycled paper" and "Dolphin-safe tuna") that became an increasingly important product feature in the late 1980s and early 1990s.

• Health-related features and natural ingredient features, such as "fat free/ cholesterol free" and "no preservatives," which have also become an increasingly important product distinction.

Fourteen Key Promotion Developments

Perhaps the most sweeping efforts by American manufacturers to obtain business are in the promotion field. Commonly perceived aspects of promotion include market research, advertising, personal selling, special sales, and general publicity. Once again, technological developments in computer and communications technologies have supported dramatic improvement. Among the specific developments emerging over the last two decades, and particularly in recent years, are these:

1. An emphasis on multiple product uses, such as Arm & Hammer Baking Soda's advertised applications for everything from cooking to cleaning.
2. A tremendous growth in the use of surveys, questionnaires, and market tests.
3. Increased levels of management participation in nonwork networks such as professional societies and charities.
4. Formal training in "active" listening for management personnel, particularly in marketing areas.
5. Marketing to aging baby boomers and "baby boom echo" offspring because of the sheer size of those market segments.
6. Establishing relationships with distributors much earlier in the life cycle of products to ensure early market acceptance through accessibility.
7. Acquisition of distribution companies with particularly suited strengths in important markets.
8. Alliances with "seed" companies that look promising and offer complementary product lines.
9. Collaborative estimating teams that include not only multiple informal disciplines but also suppliers to ensure the best possible price offering and profitability when competitively bidding for a big sale.
10. Competitive benchmarking and mapping to determine strengths and weaknesses compared to competitors', so that marketing efforts can be tailored to the strengths of one's products.
11. Development and real-time availability of comprehensive, up-to-the-minute data bases of customer buying factors.
12. Explosive use of automated information gathering about both customer buying habits and customer information sources.
13. A strong trend toward "assisting" the customer in defining his needs and in obtaining budget to make the purchase. (Do the phrases "90 days same as cash" and "Here's some information that should help you in the Capital Approval Board meeting" sound familiar?)
14. A dramatic effort toward building timely consumer causes like environmentalism and quality into marketing promotions (e.g., beer companies advertising "know when to say when")

Yes, the function of obtaining business may be the most dynamic aspect of modern manufacturing organizations. And that seems reasonable, given that it is the clearest window to the outside world—the marketplace, the arena.

The Methods Used in Defining Products/Processes

Among mainstream American manufacturers, the primary functions of product and process design are as the name implies: the development, documentation, and communication of product designs and manufacturing processes.

The Computerized Drafting Table

The methods for development of products have been most dramatically affected over the last twenty years by the evolution of computer technology. To generalize, a computer aided design workstation has replaced the drawing board. It costs a good deal more, of course, but the potential advantages are enormous. Not only will the CAD workstation draw straight lines automatically between designated points, it will fit curves and scale drawings up or down immediately before your very eyes. It will save as many versions of the drawing as you need, and effortlessly send their images to paper at the press of a button.

Consider the advances that have occurred in just this one area. First there was the paper drawing. Then there was vellum, which allowed reproduction on the old

The methods used in defining products/processes.

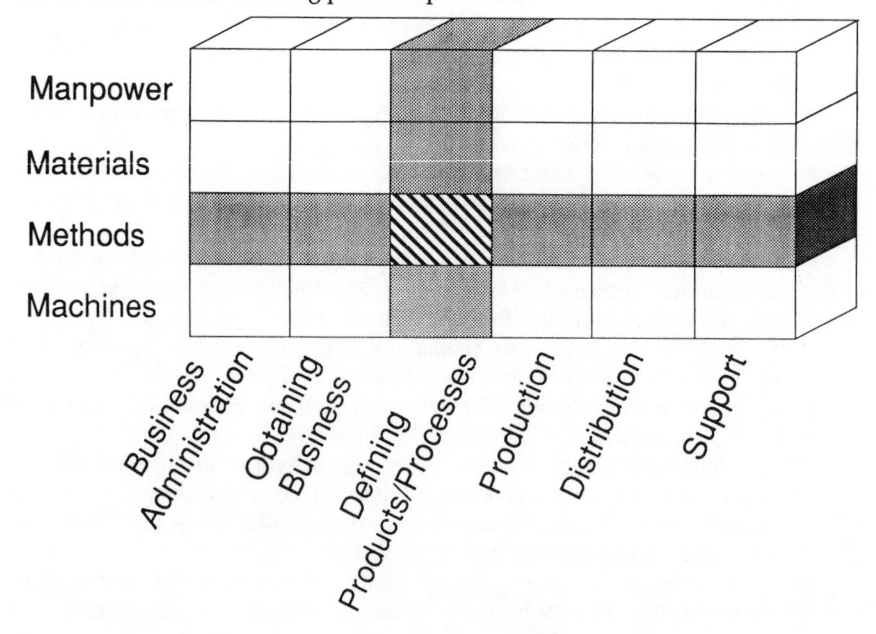

blueprint machine. Then, in the 1970s, came the two-dimensional drawings on the CRT with early CAD equipment. CAD itself evolved, allowing the depiction of "3-D wire-frame" drawings depicting all specified part edges, and three-dimensional "rotation" of the part on the screen, so that designers could view and edit the images from virtually any angle. As we move through the 1990s, we find 3-D solids modeling packages, which actually depict not only part edges but also actual surface contours, rapidly gaining market share. Coupling this capability with powerful workstation memory has brought into being a device known as the electronic development fixture, which simulates graphically the fit between parts and clearly shows interference points *before* money is spent on tooling or part fabrication. In essence, it becomes an electronic "mock-up" of the assembled product.

The use of these systems, like the MRP systems discussed in Chapter 2, didn't really become significant until the 1970s. At that point several factors contributed to an environment that would support their explosive growth, including computer hardware improvements, increased computer memory, and more capable operating systems. In addition, the demand for these tools increased significantly through the advent of complex integrated circuitry for high-tech electronics applications. The CAD system market grew from less than $25 million in 1973 to nearly $2 billion in 1983. As computer memory and speed becomes less expensive, the market and applications for these products continue to skyrocket. They can dramatically reduce the amount of drafting, documentation, design, cost estimating, and data distribution time required for design changes and new model introductions. Many companies are now struggling to make the link between computer-based product design (CAD) and the generation of usable manufacturing, scheduling, and control (CAM) plans.

An Evolving Concept of Design

In addition to the *mechanics* of the methods involved, the *concept* of how design is performed is evolving. Throughout the 1950s and well into the 1960s, the aspects known as "form, fit, and function" were the domain and guiding principles of the design engineer. The engineer was also forced to be the sole repository of knowledge in most cases for a wide range of product and process knowledge, since he bore the responsibility for virtually all aspects of downstream performance. He was to identify and incorporate all the features needed to satisfy production, serviceability, reliability, packaging, cost, esthetic, and test/inspection requirements.

During the 1970s, a major thrust—value engineering—swept through the design arena. Value analysis/value engineering (VA/VE) still has a solid, positive impact in many companies today. Like most really good ideas, its premise is relatively straightforward: minimize cost and maximize value through reduced part count and general simplification. The emphasis is cost prevention while the product is still in the design stage. By employing several structured tools and approaches, VA/VE forces participants to analyze in a disciplined and methodical way the cost elements of their designs. (One step, for example, requires the participants to assign a dollar value for each tolerance.)

Another more recent and more pervasive thrust has been toward integrating product and process design. The theory behind this movement is that the product and the processes involved in the manufacture of that product should be designed concurrently. Ideally, the manufacturing process or "producibility" person (typically a manufacturing engineer) watches over the shoulder of the product designer as he is creating. He provides valuable guidance as to the producibility of the part while the lines and surfaces are still being formulated. He thereby ensures manufacturability, ease of assembly, and minimal tooling/production cycles. Unfortunately, this is seldom easy to do, and is even less often completely practical. Several issues continue to plague these efforts, including the following:

- How do you determine that enough information exists in the CAD system to begin tool development?
- How do you determine that the design is "frozen" enough to risk tool development time and funds?
- What precisely is the product of "concurrent engineering," and how do you know when this product is complete?

These issues notwithstanding, most of the companies really making a concerted effort in this area agree that the designs yielded *are* more producible and easier to assemble than previous efforts. Progress *is* being made.

Another major design methods emphasis that grew largely out of the prodigious information storage and retrieval capabilities of the computer during the last two decades is that of group technology (GT). GT utilizes part commonality to handle parts in similar groupings, commonly referred to as part families. Part families usually are comprised of parts with similar features, dimensions, materials, and/or manufacturing process sequences. The families are usually identified and maintained by a subset of group technology known as classification and coding (C&C). C&C supports a data field that may be alpha-numerically coded with characters that define the characteristics of the parts within the specified family. The first digit in the code might designate the family as sheet metal, plastic, or glass, for example. The second digit may further define whether the sheet metal part is aluminum or steel. The third could designate whether the sheet aluminum is heat-treated. And so on. By coding and maintaining these families, manufacturers can gain great efficiencies in design because it is possible to call up the design of similar parts and merely edit one until it meets the stated specifications for the new product. It encourages part standardization and commonality and supports the procurement and production of part families, providing volume leverage to the buyer and setup/changeover advantages to the fabricator.

Mapping Customer Requirements With QFD

Finally, there is the recent impact of a movement called quality function deployment (QFD). QFD is often referred to as "deploying the voice of the customer." It has been defined by the American Supplier Institute, which offers at least one course in the subject, as follows:

> a system for translating consumer requirements into appropriate company requirements at each stage from research and product develop-

> ment to engineering and manufacturing to marketing/sales and distribution.

Basically, the QFD methodology forces the participant to clearly identify customer requirements, then maps those requirements to specific actions and targets that affect the fulfillment of those requirements. (The device used to perform this analysis is referred to as the "house of quality.") Correlations and relationships are then identified so as to determine what activities can be undertaken to most quickly and forcefully achieve the greatest customer satisfaction. It is a proven and powerful method for design engineers to identify the product characteristics most important to their end users, the consumers.

Some of the strongest proponents of this approach are in the aerospace industry. MIT's *Made in America* (Michael L. Dertouzos, et al., 1989) states,

> the commercial aircraft industry in the United States has been characterized by strong working relationships between the airplane manufacturers and the airlines. Strong customer demand for technological advances was a key factor underlying the long-term risk-taking attitude of the aircraft manufacturers.

Many QFD applications in the aerospace industry have yielded impressive results. The only real drawback with QFD is the tendency of people enamored with this tool to apply it in every situation. The old adage "When all you have is a hammer, everything looks like a nail" applies.

Process Design: The Big Picture

In terms of process design (that is, determining the best method of manufacturing a part), the current efforts of mainstream American manufacturers are centered around concepts related to "design for manufacturing" and "automated process planning."

Design for Manufacture

Design for manufacture (DFM) is commonly viewed as its major components, design for fabrication and design for assembly. Basic elements of both processes include the following:

- Material suitability analysis
- Direct labor requirements
- Indirect labor requirements
- Tooling requirements
- Energy and utility requirements
- Process capital requirements and general availability of required equipment
- Process capability (the ability of the designated equipment and methods to meet specified tolerances and to consistently produce quality parts)
- Quality verifiability (ability to inspect or otherwise verify quality levels)

Even within the major subdivisions of manufacturing there are DFM principles that have been developed and are in reasonably widespread use. In the area of machining, for example, DFM principles include

- Use of standard components (e.g., bar stock diameters)
- Utilization of work material shape to reduce machining time/effort
- Use of previous design of similarly machined jobs to minimize NC (numerical control) program development time
- Consideration of such factors as material handling and chucking during the design and material selection process
- Design within machining tolerance capabilities of existing equipment

There are other factors that relate only to design for assembly, such as the following:

- Minimization of component part count
- Part orientation
- Modularity of assembly
- Joining and fastening simplification

Automated Process Planning

Automated process planning is, of course, the automation of what has traditionally been the function of a human engineer. Process planning can be defined as the determination and documentation of which manufacturing processes (or operations) will be used in what sequence to convert material from its initial form to the form shown on the engineering model. The finished product of process planning is often called a routing, a manufacturing plan, an M.E. plan, or an operation sheet. Among the factors considered in development of the process plan are part material type, part shape, overall physical dimensions, surface finish requirements, quantity, and so on. The objectives of automating this function include reduction in the skill level requirements of the planners involved, reduced plan development time, reduced planning cost, improved plan consistency, and improved accuracy of the planning data. A typical process planning system for automated process planning may include modules for material selection, general process selection, machine selection, tool selection, fixture/handling device selection, end effector selection (for example, the "hands" of a robotic device), process sequencing, program development of NC drivers, and material optimization (for example, nesting) routines.

Variant and Generative Process Planning

There are two fundamental types of process planning in use today: variant process planning and generative process planning. Variant process planning, the most common approach, takes advantage of similarities (or common characteristics among families) of parts to develop a standard plan for the family. Then specific plans are developed for individual parts as unique requirements dictate, by retrieving and modifying the standard family plan. Among the most popular variant systems currently in use are CAPP, MIPLAN, MITURN, MIAPP, and CINTURN.

Generative process planning, the most costly and resource-consuming approach, actually synthesizes process information to create an individually tailored plan each time it is applied. Generative systems typically utilize information from engineering data bases and generate plans based on decision tree logic without the aid of human intervention. This type of approach has more stringent requirements, because the logic of process planning must be documented within the system, and the product must be clearly and digitally defined to the system. Some of the popular systems of this type currently marketed include AUTOPLAN, AUTAP, APPAS, and CPPP.

The Methods Used in Production

The methods utilized in our current production environments may be viewed as falling under two broad categories: production management and actual conversion methods. Production management typically consists of production planning/control, quality assurance, and shop floor supervision activities.

Production Planning and Control

Production planning and control moved through the fundamental steps of MRP (material requirements planning), MRP II, and just-in-time manufacturing

The methods used in production.

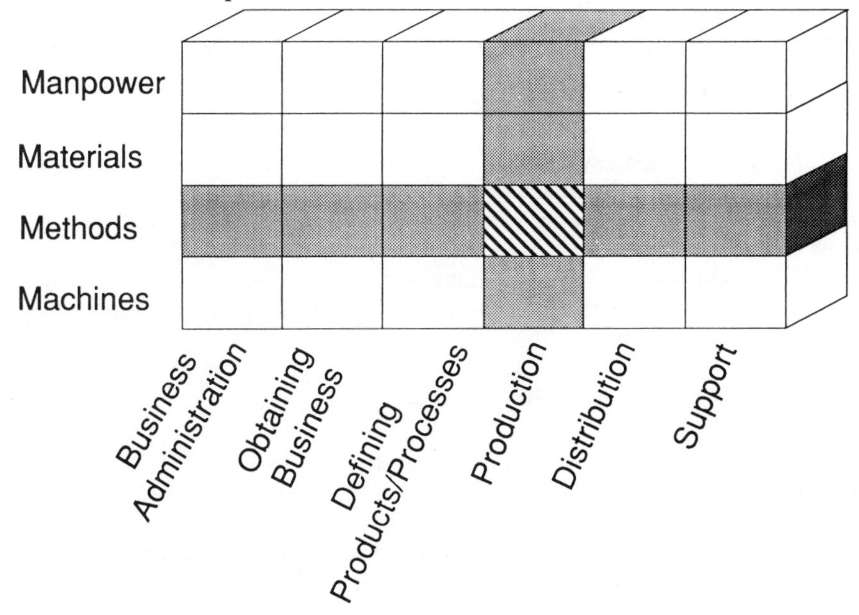

between the 1960s and 1990. These were all fairly basic shifts in manufacturing philosophy, although there is a parent-child relationship between MRP and MRP II. (Although MRP II came later, it is the parent in the current MRP II environment; this evolutionary process was discussed in Chapter 2.) Basically, the salient points for this discussion are as follows:

- MRP (material requirements planning) is a computer-based system of netting demand for finished products against existing finished goods inventory, work in process, raw materials, and open purchase orders at each level of a structured bill of materials. MRP then generates planned orders for production and purchasing, for the actual time each item/material is needed, according to lead time offsets designated within the system.
- MRP II (manufacturing resource planning) is a much more global application of the MRP principles. It begins with top-level production planning ("How many widgets are we going to sell this year?"), drives through capacity planning, material requirements planning, purchasing, and shop floor production ordering and reporting, and finally "closes the loop" in each step by providing feedback to parent systems and visibility of exceptions to managers for their attention and resolution. MRP II is also a computer-based system or series of systems (generally referred to as "modules" by software vendors).
- JIT (just-in-time) is not a computer-based system like MRP and MRP II. It is a philosophy advocating elimination of waste in all its forms throughout a company's operations. It focuses on speed, quality, line-of-sight management, and cooperative relationships between suppliers, producers, and customers.

Rather than static schedules and a production order based on due dates, just-in-time production systems often employ a pull system, which prioritizes all work on a first-in, first-out basis combined with synchronized production rates. It works to minimize inventory and throughput time rather than optimize production runs to compensate for long changeover periods.

The best current method is probably a combination of MRP II and JIT. MRP II is almost certainly the best production planning method devised to date. However, it often loses control on the shop floor because it simply isn't dynamic enough to keep up in "real time" with daily production circumstances. JIT offers positive dynamic production control with minimum inventories, but does nothing for planning. Blended properly, these techniques offer a comprehensive and effective production planning and control system.

Quality Assurance: A Shift Toward Prevention

Quality assurance (QA) is another area that has developed significantly over the past three decades. Originally, QA was an inspection-based discipline whose job it was to identify defects and assign blame. The focus really began to shift away from inspection and assignment of blame toward a "root cause" and defect *prevention* orientation in the 1980s.

Inspection methods involved the sampling of raw material, parts, and assemblies at various stages of completion to determine whether they met design specifications. Generally, everything sampled that was rejected fell into one of three categories, depending upon the severity of the problem: rework, scrap, or use as is. The actual inspection methods varied (and continue to vary) according to the material and manufacturing processes involved. Dimensions, for example, are often checked by a coordinate measuring machine (CMM), micrometer, or simple linear scale. Hardness is measured by Rockwell tests, spark tests, and various forms of destructive testing. Porosity may be identified by ultrasonic scanning, and impurities by use of an X-ray machine.

As it became apparent that we needed to drive upstream from defects to identify and correct "root causes," new tools related to structured problem solving and experimentation came into widespread use. Among the most important of these are the Ishikawa fishbone diagram, and the design of experiments (DOE). As mentioned, most recently the emphasis has been shifting away even from root cause (although it is clearly recognized that driving to root cause will be required as long as there are defects) and toward defect prevention. In defect prevention, even more tools have emerged, such as statistical process control (SPC) and failsafing. All together, there is now an impressive array of analytical quality tools available to the American quality engineer and to management in general, including cause-and-effect diagrams, histograms, flow charts, scatter diagrams, Pareto charts, run charts, and process control charts.

The individuals who have most heavily influenced our thinking on the topic of quality today include the following:

- *Dr. W. Edwards Deming,* who preached the gospel of variation reduction to consistently improve productivity
- *Joseph M. Juran,* among the first to deal with the broad management aspects of quality, who believes in structured improvement initiatives, a deliberate sense of urgency, extensive training, and strong upper-level management leadership
- *Philip B. Crosby,* who describes quality as conformance to requirements and insists that "quality is free"
- *Dr. Armand V. Feigenbaum,* the originator of "total quality control," who emphasizes quality as a business strategy and a primary determinant of competitive strength
- *Dr. Genichi Taguchi,* whose most significant contribution is probably the "loss function," a method for defining quality as the characteristic of a product that avoids loss to society between shipment and the end of the product's useful life

All of these gentlemen have made substantial contributions, but the pioneer and standard-bearer of "the faith" in this area was Deming. The concept of continually reducing variation underlies almost all of our modern quality philosophy, including recent success stories like Motorola's Six Sigma program.

Shop Floor Supervision: From Dictators to Participators

Shop floor supervision methods have, of course, changed with the times as well. Even as late as the 1960s and early 1970s, shop floor supervisors were often chosen for their physical stature and booming voice. Crowd control and the ability to lead by intimidation were often the hallmarks of the shop foreman. Labor unions and collective bargaining, although making tremendous strides for blue-collar workers, served in many cases to drive a wedge between management and workers. The reforms were long overdue and in most cases richly deserved, but they changed a relatively efficient (if many times dictatorial) management style into one that can be described fairly only as laissez-faire. The role of supervisor was often reduced to a frustrating array of daily battles with union stewards, committeemen, and officials. It took significant time out of both employees' and supervisors' workdays and dramatically restricted the deployment and cross-training of workers.

With the scientific methods of work measurement introduced by Frederick W. Taylor in the late nineteenth century came the field of industrial engineering and work (incentive) "standards." Combined with the later advent of computerized labor reporting, shop floor supervision began, particularly in the 1970s, to manage "by the numbers." The computer made it possible to measure direct labor productivity weekly, then daily, and now on a virtually "real time" basis, even from remote locations. And measure we did. Soon, almost immediately in fact, direct labor became the central focus of management.

There were several problems with this approach, three of which were (and often still are) paramount:

1. It quickly generates a "we versus they" relationship between management and worker.
2. It concentrates on the least-cost slice of the cost of goods sold.
3. It tends to focus management attention on short-term gains rather than exposing root causes of performance problems and preventing them.

Shop floor supervision and general shop floor management practices have been far more evolutionary than revolutionary. It is fair to say that mainstream American manufacturers appear to be moving from a fairly authoritative Theory X type of management to a more participative Theory Y management style. We see phenomena such as "quality circles," "natural work groups," "problem-solving teams," and other forms of interdisciplinary teams taking an ever-increasing role in their own management in virtually every major U.S. manufacturing organization.

This means that more training time is required per employee, and that employees must be better skilled interpersonally. They must be able to not only perform their own jobs but also be alert to opportunities for improvement. They need problem-solving skills and good communication and idea presentation skills as well. The days are quickly disappearing when an employee "checked his brain at the gate" and all one needed to make an adequate living in the United States was the will to work and an able body.

Cycle Time Reduction: Productivity's Biggest Boost

In terms of general productivity improvement and the ability to continuously improve quality while reducing cost, there is one activity that has probably overshadowed all others in recent manufacturing history. It is the relentless pursuit of something called "cycle time reduction."

As mentioned in Chapter 2, the reduction of throughput time in every area of the manufacturing enterprise from marketing through distribution and customer support is tantamount to competitive strength. Nowhere is that more true than in actual shop floor manufacturing activities. Manufacturers who consistently measure, focus on, and reward reductions in throughput time on the shop floor virtually always remain competitive and improve their overall productivity. Cycle reduction forces manufacturers to constantly reduce "non-value-added" activity and its related cost. That means improved quality, less overhead, and greater employee commitment. It has probably been the single most powerful initiative in the effort to revitalize American industry. There are two primary ways that cycle time reductions are achieved in the manufacturing setting, depending on whether the setting is an assembly or a fabrication operation.

In assembly environments, manufacturers have discovered that the secret of reducing throughput (cycle) time is continuously manning and reducing time required on the "critical path." The critical path in any assembly process is the sequence of jobs that requires the longest time to complete. In any assembly program, there are many individual assembly operations. Each operation cannot be started until one or more preceding operations upon which it depends are completed. That operation, in turn, must be completed before one or more downstream operations may be started.

Because operations require different amounts of time to complete, the longest "string" of operations is seldom obvious. It also often changes as problems are experienced or work is redistributed. Therefore, continuously manning the critical path is essential to producing assemblies as quickly as possible. Improvement efforts on operations other than critical path operations are largely wasted, since they will not reduce overall throughput times.

In fabrication and "process industry" environments, the approach that American manufacturers have found most useful is continuous reduction of interoperation queues. In virtually every fabrication and process manufacturing operation, the lion's share of throughput time (typically more than 75 percent) is consumed by products (or product components) just sitting between operations. Eliminating the reason the product sits there, sometimes referred to as "the cause for the pause," is an extremely powerful way to improve productivity of not only direct labor but also material. (Material costs are almost always several times greater than direct labor in terms of the portion of cost of goods sold represented. Generally, material represents more than 50 percent of COGS, while labor is in the neighborhood of 10 percent.)

The pendulum also appears to be swinging back somewhat from the extraordinarily complex machinations of a computer integrated manufacturing (CIM) philosophy toward a more simplified view typified by terms like *line-of-sight management* and *pull systems*. This is in keeping with the increasingly participative management style and with certain practices associated with just-in-time and total quality management.

The Methods Used in Conversion Activities

The methods involved in actual conversion activities are, of course, very diverse. They are most often viewed in aggregate in categories related to the type of materials they are most commonly associated with. This suits our purposes, especially since the relationships between materials and methods are part of our focus of study here. However, a word of caution: Don't let existing relationships unnecessarily confine your thinking about potential applications of methods and/or equipment to specific materials. (Methods that were applied only in the metals world for decades now are widely applied in plastics, for example.) Consider, then, this list. It is by no means exhaustive, but it will serve to provide a general picture of the methods currently employed in production by mainstream American manufacturers as we pass the midpoint of the 1990s:

I. *Metal-Forming Methods*

- Stamping
- Brake forming
- Roll forming
- Section contour forming
- Metal spinning
- Deep drawing
- Rotary swaging and hammering
- Wire forming

II. *Metalworking and Metal-Forging Methods*

- Die forging
- Hot upsetting
- Die rolling
- Hot extrusion
- Cold extrusion
- Impact extrusion
- Cold drawing
- Cold heading
- Thread and form rolling

III. *Metal Removal Processes*

- Flame cutting
- Contour sawing
- Planing
- Shaping
- Slotting
- Turret lathe machining
- Bar and chucking machining
- Milling
- Drilling and boring
- Hobbing
- Broaching
- Gear cutting and shaping
- Belt grinding
- Production grinding (e.g., cylindrical, surface, centerless)
- Tumbling and rotofinishing
- Honing
- Lapping
- Superfinishing

IV. *Casting Methods*

- Sand casting
- Permanent mold casting
- Centrifugal casting
- Die casting
- Plaster mold casting
- Investment casting

V. *General Molding Methods*

- Powder metallurgy
- Plastics molding
- Rubber molding
- Ceramics molding

VI. *Treating Methods*

- Heat treating
- Shot peening
- Organic finishing
- Marking

VII. *Plastics and Composites Manufacturing Processes*

- Compression molding
- Transfer molding
- Injection molding
- Extrusion
- Rotational molding
- Blow molding
- Thermoforming
- Reaction injection molding
- Casting
- Forging
- Foam molding
- Reinforced plastic molding
- Vacuum molding
- Pultrusion
- Filament winding
- Tape laying
- Resin transfer molding
- Matched die molding
- Calendering
- Film casting

VIII. *Assembly*

- Traditional welding
- Spin welding
- Solvent welding
- Ultrasonic welding
- Heat sealing
- Ultrasonic staking
- Spot welding
- Seam welding
- Projection welding
- Butt welding
- Soldering
- Brazing
- Adhesive bonding
- Mechanical assembly

Even in the mid-1990s, a great deal of our fabrication is still traditional metal forming and metal removal. This situation is changing as our products are "demassified," made recyclable, and increasingly made from nonmetallic substances. Nonetheless, we will see the preponderance of these techniques sustained through the balance of this decade.

The Methods Used in Distribution

The methods utilized in our current distribution environment are fairly categorized in terms of internal stocking and controls, logistics, and distribution center management. These functional areas, traditionally regarded as the distribution division, account for a sizable portion of the cost of manufacturing and selling goods in the United States. In fact, by 1982 the portion of the U.S. GNP that could be attributed to "logistics" amounted to about 23 percent, or $680 billion. For many companies, it was their second-highest cost element. Of that cost, transportation was the largest element (about 44 percent), warehousing was the next largest (about 27 percent), inventory carrying cost was next (around 25 percent), followed by general distribution administration (around 4 percent).

Internal Stocking and Controls

Internal stocking and controls is tightly coupled to sales forecasting, and as such it requires market intelligence and other data typically held in the marketing

The methods used in distribution.

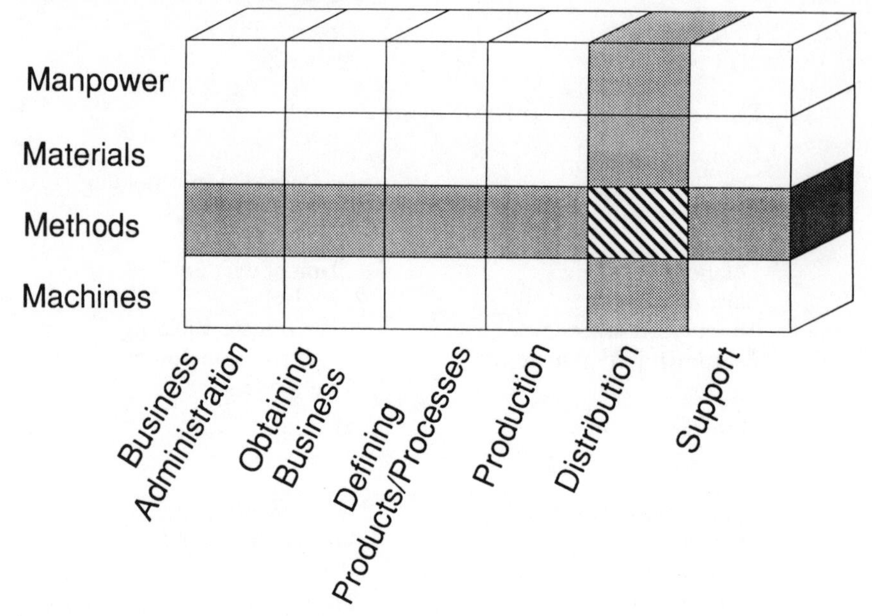

and general administration areas. This has traditionally been a problem for distribution management personnel, because they have been whipsawed between late and limited receipt of information critical to adequate planning of their operations and the need to react quickly to dynamic demands from the customer base. In general, it is probably fair to say that those companies with the best market "pulse" reading and strategic planning capabilities are also among the best at distribution planning.

As U.S. manufacturers moved from MRP to MRP II (discussed in Chapter 2), a module of MRP II, called distribution requirements planning (DRP), was developed. The real emphasis of DRP was on taking advantage of the information available to MRP and providing both forecasted and actual sales information to distribution managers. In addition, DRP provided visibility of inventory levels, work in process, and other information valuable for distribution planning and decision making. Put simply, it allowed everyone from sales through distribution to work with the same set of numbers. It also "closed the loop," allowing distribution centers to feed back inventory depletion numbers and inventory turnover information on a continuous, "real time" basis. Salespeople could, for the first time, give their customers a fairly accurate date of availability for their goods based on knowledge about whether goods were available in warehouse locations or were somewhere in process in the factory itself. Most of the integrated systems development work associated with major DRP efforts occurred in the 1970s.

Logistics

Logistics activities include route planning, fleet management, and general finished goods inventory distribution/redistribution. Most of these functions, and the methods associated with them, are not substantially different from what they were two decades ago. There are some differences related primarily to the efficiency of transportation systems, such as the advent in the late 1950s and continuing throughout the 1960s of the modernized American highway system. In addition, transportation has become generally more efficient from the standpoint of fuel required per mile. However, transportation costs associated with fuel (cost per gallon), insurance, and maintenance labor have also risen dramatically over that same period.

The Overnight Delivery Boom

One of the most interesting developments in the area of distribution over the last decade has been the emergence of third-party distributors of virtually everything. One major contributor to this phenomenon is the overnight delivery boom. Companies like Federal Express, Emery, and Flying Tigers have changed the way customers view the availability of goods and their overall expectations of service. As a result, many companies have adopted (some voluntarily) special expedited delivery services through their own distribution networks.

Still others, recognizing that they cannot competitively provide these rapid distribution and delivery services, have aligned themselves strategically with one of the carriers that specializes in that business. A few years ago, Federal Express spawned a new division known as Business Logistics Services to provide just this kind of support. Thus far, it has proved very successful indeed.

The kinds of warehousing services provided in these scenarios include centralized hub warehousing, flexible storage and handling, sorting, kitting (assembling kits of components such as spare parts to be sold in one package), even final product configuration. These distribution and delivery services will perform inbound and outbound quality inspection, material and order tracking, and product withdrawal/recall management. In terms of transportation, they will do traditional inbound/outbound goods transportation; integrate multimodal transportation; perform just-in-time inventory management, same-day delivery, next-day delivery, best-way delivery; and support private transportation networks.

Distribution Center Management: Robots Manning the Warehouse

Distribution center management methods have dramatically evolved over the last two decades as a result of computerization, robotics, and electronic data interchange (EDI). Again we see the impact of computerized inventory management systems, such as the inventory control module of closed-loop MRP systems. Almost without exception, all inventory is now "on-line," with all storage, picking, issuing, and kitting computer-controlled. Most major distribution operations now utilize bar coding and remote (often portable) readers ("scanners"). Other methods employed include magnetic stripe readers (similar to the ones used by automatic teller machines to read the magnetic strip on the back of your plastic bank card) and optical character readers (OCRs). In many cases, mainstream warehousing operations have employed robots to store and retrieve their goods, calling them automatic storage and retrieval systems (AS/RSs).

Finally, centralized distribution managers currently utilize EDI to view the sales trends in various geographic regions, evaluate current and planned inventory distribution among their regional distribution centers, and reallocate finished goods in reaction to and in anticipation of customer buying activity.

The Methods Used in Product and Customer Support

The methods used by mainstream American manufacturers to support their customers after products have been delivered include delivery of repair parts, field service of the product, return of defective products, warranty claim management, production and delivery of technical publications/instructions, and collection and management of field performance and field failure information.

Toward Faster Part Repair

Response time associated with repair part delivery has proved to be an increasingly important differentiator of companies in the minds of customers. The ability to quickly obtain a critically needed replacement part is sometimes one of the most important factors in establishing and maintaining customer loyalty. In many cases, time literally means money to the customer. Broken vending machines mean lost revenue. Broken harvesting equipment means less crop harvested and, again, lost revenue. Broken-down trucks mean reloading of goods, late deliveries, spoilage, and, again, lost revenue. Customers want their equipment to perform without failure, and when it does fail, they want the replacement/repair parts fast.

The methods used in product and customer support.

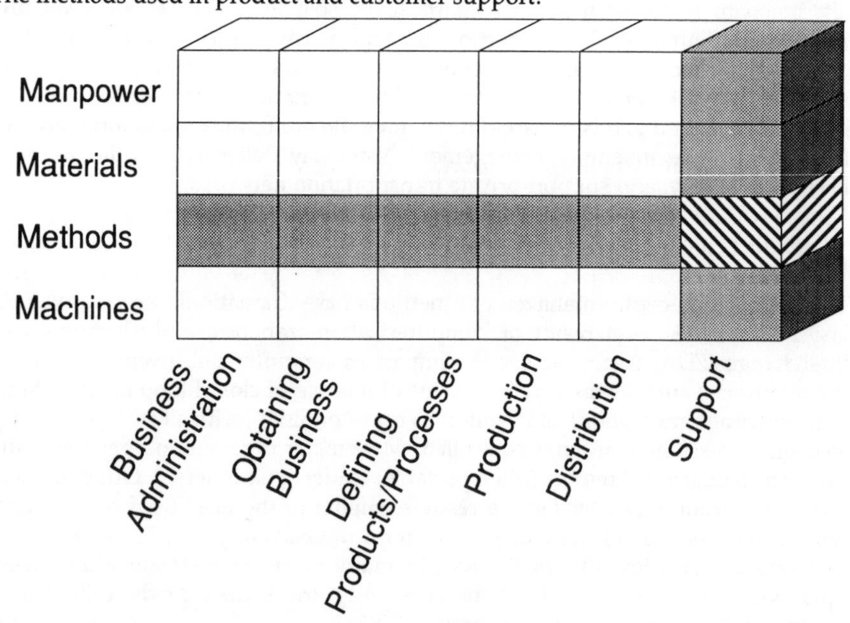

(We are reminded of the discussion in the previous section of this chapter about third-party deliverers of goods and their same-day and next-day services.) And of course there is an EDI application here. Many customers are now in possession of terminals that allow them to input their orders for both goods and repair parts directly to the factory.

Where Repair Service Needs Fixing

Over the last decade, it has become increasingly difficult to do good up-front field service, for several reasons:

- The recent emphasis on shortened cycle-to-market time means that there is less time for training repair people and that designs are far more subject to revision.
- Products are far more complex and compact. Many of the best-designed products are modular, allowing relatively quick and easy replacement of defective modules. However, not all products are designed in a modular manner (indeed, not all products lend themselves well to modular design).
- The diagnostic equipment required to determine which part of the product is defective or damaged is increasingly expensive and complex (the dark side of computerization).
- Labor associated with the time of repair people who must be more highly educated, more highly trained, and more technically equipped is also increasingly expensive.

Mainstream manufacturers, in their struggle with these circumstances, are integrating customer support people, particularly repair people, into concurrent engineering teams to get them involved with the product design and to make them familiar with the product as soon as possible in its life cycle. They are also training repair people by making them assemblers of the product for the first few units, so that they have an intimate understanding of the assembly/disassembly process.

Many manufacturers have adopted a policy of strategically aligning themselves with third-party repair companies that specialize in the repair of a particular type of product. These relationships can have real bonuses in terms of leveraging the expensive equipment and repair facilities of others and in terms of gathering important competitor intelligence. However, there are also potential risks. It is more difficult to ensure responsiveness to customers, since the manufacturer has limited control of the personnel actually performing the work.

Return of the defective products is often being addressed in the same ways. Third-party services, such as Federal Express's Business Logistic Services, offer equipment exchange and upgrade services, product/component repair and return services, and same-day service for some products. Once again, American manufacturers have found over the last decade that customer service and customer satisfaction are extremely important to customer loyalty. Having loyal customers has proved to be a far less expensive path to market share than obtaining new customers.

Providing service for a small fee is also a way manufacturers are turning customer service into a profit center. Warranty claim management appears to have been strongly affected during its evolution over the last two decades by the concepts of insurance and preventive maintenance. A growing percentage of the revenues contributed by customer service departments is a result of the now famous (infamous?) "service agreement" or "maintenance agreement." Basically, this works on the same principle as insurance. If a lot of people pay a little money for a maintenance agreement they never take advantage of, a few people can execute a claim against their agreements, and there is still considerable profit left over for the manufacturer. In addition, the maintenance schedules now provided with many new products (automobiles, for example) outline a regimen that can only be described as advantageous to the profitability of the service center.

Production and delivery of technical publications has also become a more difficult process with the advent of time-based product development and delivery and the attendant emphasis on cycle-to-market. Most of the reasons are the same as those just listed regarding field service. Technology advances have recently provided some offsetting advantages in this area, however. Desktop publishing, the ability to efficiently merge text and graphics, the development of training videos, and computer-based design have made technical publication production methods much faster and easier.

Finally, the methods used in gathering, analyzing, and distributing field performance data have improved dramatically with the advent of computerization and electronic data interchange. EDI supports immediate feedback about the nature, volume, geographic location, and time associated with field performance problems. The data can be far more quickly assimilated and analyzed by customer service and design engineering managers and technicians. In addition, the computer's number-crunching power has put powerful analytical tools into the hands of almost everyone who uses field service data. The result of these developments has been general improvement in the timeliness and response to field performance problems to a degree previously unheard of.

It thus seems fair to say that our customer and product support organizations are currently driving toward profitability through automation and quicker response times. In many cases, these focused efforts won't be nearly enough to make this area successful in the next few years, but broader, more fundamental improvements are in store. We'll discuss these in subsequent chapters of this text.

Chapter 6

Dominant Materials in the Information Age

As we move into the last half of the 1990s, it is important to grasp that "materials" as they are employed by manufacturing organizations (see Exhibit 6-1) encompass far more than metal and plastic. We have discovered that it is becoming vital to include data, information, and knowledge in our concept of materials.

The preface of what in my opinion is one of the most important books to be published in recent years, *The Materials Revolution* (MIT Press, 1988), by Tom Forester, begins with these words:

> Three megatechnologies will dominate the last decade of the twentieth century: information technology, biotechnology, and new materials. The Japanese recognized this years ago—that's why they targeted all three for special R&D effort. Only now is the United States beginning to wake up to what's been going on.

Unfortunately, Forester couldn't be more correct. In looking at the materials currently in use throughout manufacturing organizations in the United States, I initially thought that the significant changes were all occurring only on the factory floor with such items as composites and ceramics. Having thought through the "material" used in each area of our model (for instance, business administration, marketing/sales, design of processes and products), however, I realized that the "material" of these other areas is primarily data. That data *has* changed. It is fundamentally different in many cases from what it has ever been before. And in fact, data and the distilled version of data called information are increasingly built in to not only the products but the processes supporting actual production activities.

Therefore, we are faced in the United States with a shift in materials not only away from traditional production materials (metals) to newer, lighter, stronger successors (such as ceramics and composites) but away from paper-based history toward electronic projections, simulations, trend analyses, and heuristics. We are moving away from autopsies, if you will, and toward preventive medicine. The new materials used in these areas include analytical and management software of every imaginable type, real-time analyses of massive quantities of transactions, from sales to distribution, and new perspectives of the "big picture" made available to managers for the first time through unprecedented levels of systems integration.

Both the shift in production materials and the shift from paper records to electronic projections have already had dramatic effects on American manufactur-

Exhibit 6-1. Dominant U.S. manufacturing materials.

U.S. Manufacturing Materials

Business Administration

- Foreign Market Analytical Data
- Competitor Data
- Stock Ownership Data
- Domestic Market Analytical Data
- GAAP Updates
- Cash Flow Data
- Inventory Data
- "Cost-of-Quality" and Other Cost Data
- ROI, ROS, ROA
- Utilization Data
- Demand Data
- Schedule Data
- Financial Analysis Software
- Modeling Software
- Asset Management Information

Obtaining Business

- Foreign Market Analytical Data
- Competitor Data
- Domestic Market Analytical Data
- Actual Sales History
- Projected Sales
- Customer Records
- Cycle-to-Market Data
- Customer Surveys
- Distributor Data
- Brochures
- Media Exposure
- Modeling Software

Define Products and Processes

- 3-D Models
- Performance Data
- Analytical Software
- "Customer Needs" Data
- Producibility Data
- GT Data
- Materials Capability Data
- R&D Information
- Simulation Software
- Standard Parts Data

Production

- Schedules
- Inventory Data
- Requirements Data
- BOMs
- Routings
- Standards
- Labor Accounting Data
- QA Information
- QA Performance Data
- Vendor Performance Data
- Cycle Time Data
- Training Materials
- Cost Data
- PM Data
- Metals
- Polymers
- Ceramics
- Composites
- Semiconductors
- Construction Materials
- Timber
- Fiber
- Paper
- Glass

Distribution

- Delivery Schedules
- Market/Sales Forecasts
- Route Data
- Transportation Cost Data
- 3rd Party Data
- Field Performance Data
- Inventory Data
- Delivery Performance Data
- Bar Code Schema
- Inventory Management Software
- Order Tracking Software
- Carrier Cost Performance Data
- Product Design Information

Support

- Repair Part Sales Information
- Repair Response Time Information
- PM Data for Repair Equipment
- Repair Crew Management Information
- Repair Performance Management Information
  - Cost
  - Time
  - Quality
- Service Agreement Management Information
- Warranty Management Information
- Information for Technical Publications
- Drawings, Specifications
- Training Videos
- Training Modules
- Performance Management Software

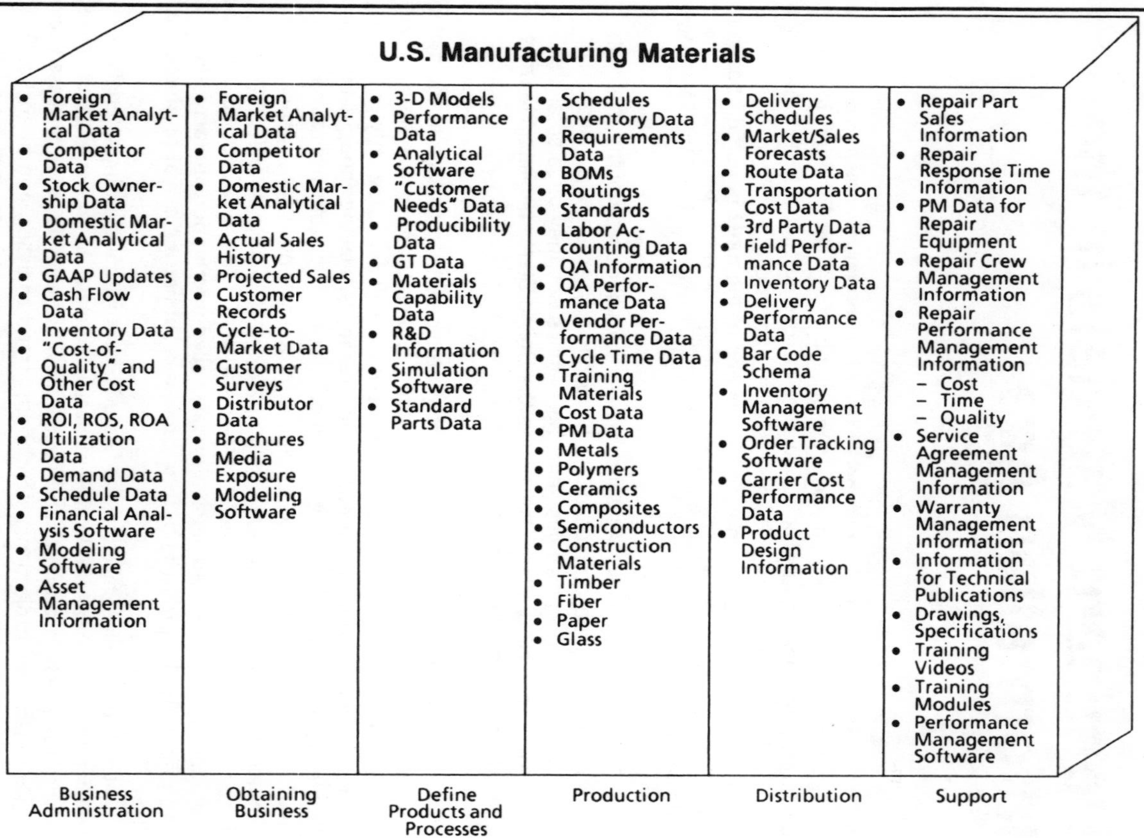

ing operations and the products of our factories. Some of these consequences are reviewed in this chapter and others in subsequent chapters.

How is this information age material different from what American manufacturers have traditionally used? How and why has it changed over the last thirty years? The answers lie in computers, electronic data interchange, and software.

The data we use today is much broader in scope. It can be gleaned from anywhere in the world over a telephone line any time of day or night. These massive volumes of available data would be completely overwhelming but for the amazing ability of computers to ingest, analyze, and convert the vast seas of data into usable information.

The data we use today is far more timely. The speed of electronic data interchange (EDI) data is virtually the speed of electricity, and in the case of fiber optically moved data, it is literally the speed of light. No more waiting, in most cases, for weekly or monthly reports. It's all there, up to the minute, all the time.

Our data today is also more useful. It can be analyzed and converted automatically by linear equation or any other of scores of approaches into projections of upcoming events. It can be displayed graphically or textually or both, and distributed again in its most potent and useful form electronically in a matter of seconds or minutes. It can be screened electronically and tailored to its user. In general, data and information are more plentiful, more available, and easier to use.

The Materials Used in Business Administration

The materials used in business administration activities as American manufacturers traverse the 1990s is of two fundamental types: data and software. The data

The materials used in business administration.

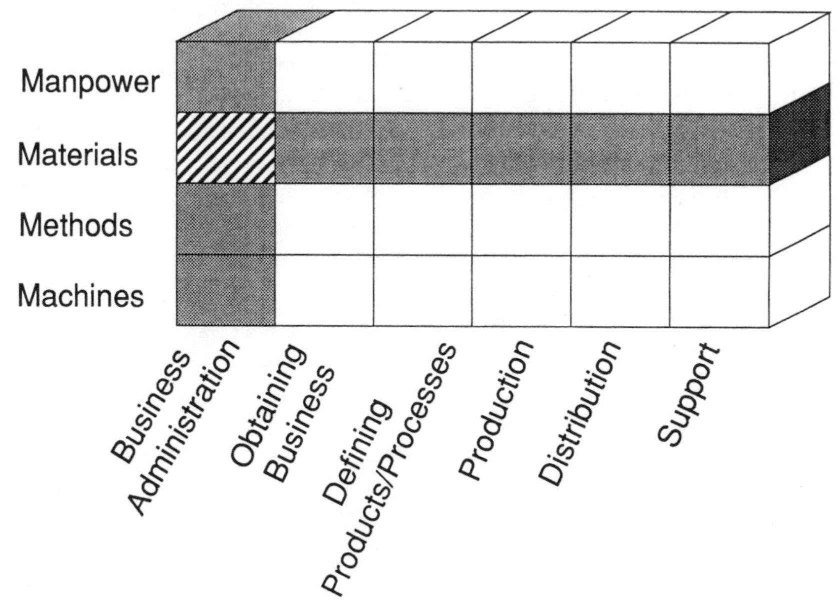

required for proper administration of business administration activities such as accounting, strategic planning, human resources management, and change management includes:

- *Market data*—both foreign and domestic, and in most cases the mix is increasingly dominated by foreign interests. Sales information, market trends by geographical locale, market stratification data and buying trends by age group, cultural orientation, and gender. Constant market share analyses.
- *Company stock ownership data*—both historical and current. In an environment increasingly characterized by takeovers, mergers, acquisitions, and divestitures, it is vital that management be aware of company holdings and significant movements in that area.
- *Generally Accepted Accounting Principles (GAAP) updates*—to ensure compliance with all federal tax regulations. The constantly changing financial reporting requirements make it difficult to ensure the accounting records are maintained in accordance with Generally Accepted Accounting Principles and other regulatory standards (for example, state or federal government requirements).
- *Normal financial data*—all of the elements contained in normal cash flow analyses, and ROI/ROA/ROS (return on investment/return on assets/return on sales) calculations: sales, assets, inventory, COGS, interest, depreciation, and so on.
- *Detailed cost management data*—including some relatively new categories such as the "cost of quality." Now more than ever, American manufacturers are feeling the need to reduce and tightly control costs. With the advent of new methods such as activity based costing (ABC), different data is required, and more of it, than ever before.
- *Asset utilization data*—for capital assets such as plants and major equipment, primarily. A major emphasis in many downsizing American manufacturers over the last decade has been on reducing their "break-even point" (the lowest level of capacity at which the factory can be utilized and still be profitable).
- *Financial analysis software*—for stockholders' financial report development and recording/reporting earnings as well as tax liability to federal, state, and local governments.
- *Modeling software*—for what-if scenario building, strategic planning, and determining everything from optimum plant locations to which computer to buy.

The Materials Used in Obtaining Business

The materials used in obtaining business are not greatly different from those used to perform business administration functions in many cases. Marketing and sales personnel are also quite interested in buying patterns and other market-related intelligence. The primary difference in market intelligence relates to the way the data is analyzed and used. The objectives of marketing are not related so much to strategic planning and development of corporate policy as they are to anticipating market opportunities and exploiting them. Therefore, while many of the data

The materials used in obtaining business.

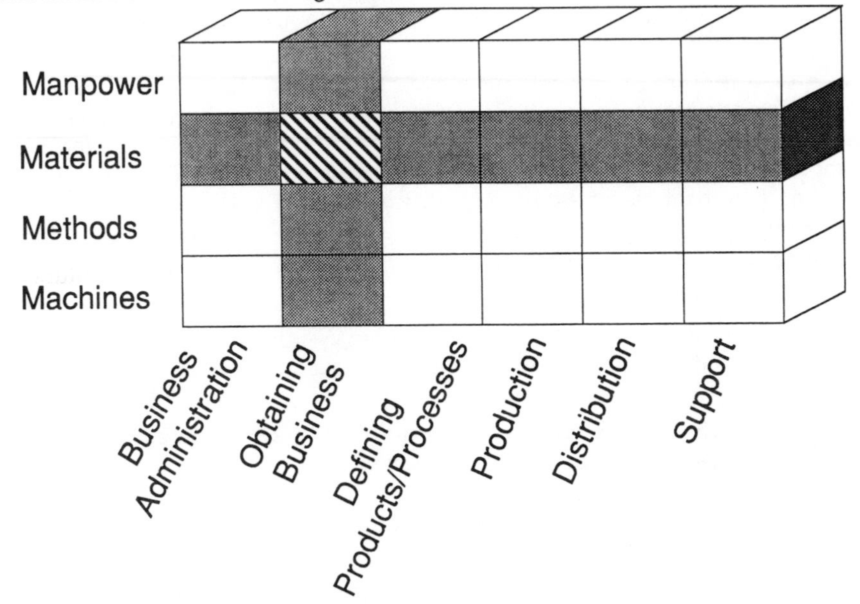

elements are common to other business functions, the analytical software utilized often differs significantly.

The data and software materials used by the marketing group, then, generally include these:

- *Market data*—both foreign and domestic. Again, in most cases the mix is increasingly foreign. Sales information, trade studies, market trends by geographical locale, market stratification data and buying trends by age group, cultural orientation, and gender. Constant market share analyses.
- *Competitor data*—not only traditional market share data, but (especially toward the end of the 1980s) looking particularly closely at competitors' special strengths (something called "best practices") and weaknesses. What was commonly considered some years ago to be industrial espionage is now often regarded as "competitive benchmarking."
- *Sales performance data*—both historical and current. Anyone familiar with "charting" specific stock market issues to determine and analyze correlations (known as betas) will understand the value of this information. It is often possible and rewarding to identify correlations between sales and external influences (such as independent indices, regional economic trends, and so forth).
- *Customer records*—for past and current as well as potential customers. Different data is relevant depending on the product and the market, but every sales organization requires such data. The manner in which this kind of data is gathered has undergone dramatic change, with checkout registers directly linked to order entry devices in some retail outlets and sales volume on many consumer products classified by zip code.

• *Cycle-to-market data*—showing historical performance and defining current capability to turn out and deliver product in a time-effective manner following order entry.

• *Customer survey data*—both preferences and needs. The customer survey has existed for decades, but today there is new emphasis on not only listening to the customer but "deploying the customer's voice" throughout the manufacturing company. Formally known as quality function deployment (QFD), the process requires that data be gathered and structured in a solid, quantifiable, and usable way.

• *Distribution data*—to be able to quickly identify the location of finished goods at any distribution location and allocate them to a pending sale.

• *Brochures and marketing materials*—requiring design data and specifications as well as the ability to merge text and graphics into attractive, "customer friendly" documentation (desktop publishing).

• *Advertising rate and vehicle information*—by geographical location. Not only what is available at what cost, but how effective each vehicle has been at increasing sales.

• *Modeling software*—analytical tools that analyze and present (both textually and graphically) the results of marketing expenditure permutations and identify the combination of activities and locations likely to yield the greatest sales impact.

The Materials Used in Defining Products/Processes

The materials used in product and process definition include data from a wide variety of sources, including customers, trade studies, technical journals, textbooks, and supplier parts catalogs. Although many of these items are available on-line, many others are not (yet). Therefore, most of the required data continues to be paper based. Software used in this area is typically much more complex than the analytical software used in other divisions of the manufacturing company and requires far more technical expertise to use. It is generally the least "user friendly" and the most specialized.

The material used in defining products and processes includes the following:

• *Assembly and component models*—varied in type from mechanical structures to chemical chain configurations. In the more modern design environments where hard goods are produced, computer-based design systems allow all designs to be digitized and electronically stored on magnetic media (tapes and disks). The models can be shown as wire frame or surfaced in multiple colors. They can be rotated in three dimensions and viewed from any angle on-line before they are ever manufactured. However, some American manufacturers still have not significantly invested in this technology and rely on paper-based designs.

• *Performance data*—field and laboratory test performance data, as well as actual customer-reported performance numbers. Again, with the advent of EDI and the miniaturization of performance-reporting instrumentation, this data is now often available in unprecedented volume and with incredible speed.

The materials used in defining products/processes.

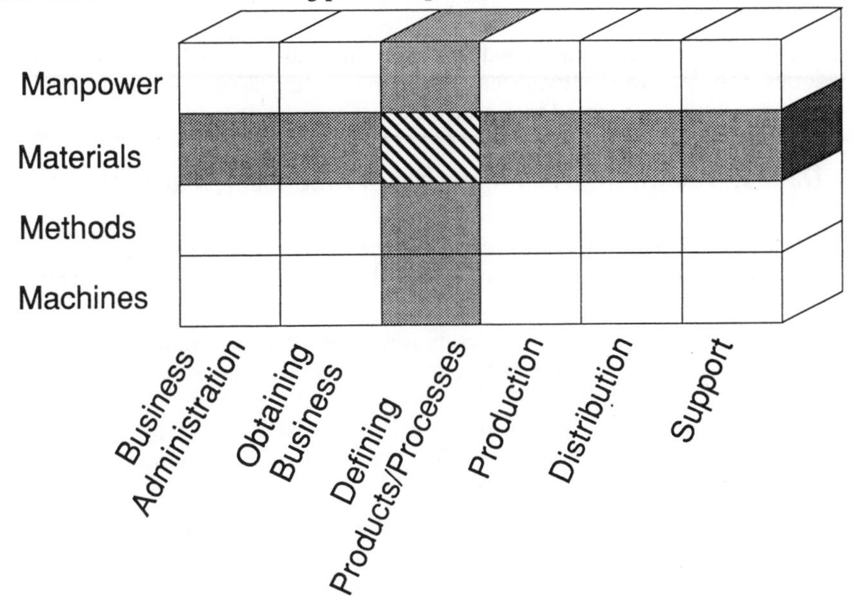

• *Group technology data*—especially available parts by classification. Classification coding of parts by characteristic, as described earlier in Chapter 5, has made available to the design engineer a wealth of design time-savers and offers cost-cutting opportunities through commonality of components and hardware. With immediate visibility of families of parts with similar characteristics, there is less tendency on the part of designers to "reinvent the wheel" each time a product is conceived or revised.

• *Materials capability data*—often available through industry technical publications. This data provides listings of likely applications, material usage limits, and even historical cost figures both for initial procurement and downstream usage on the basis of fatigue and reliability.

• *Standard parts data*—most often retrieved from supplier parts catalogs. Lists general specifications, characteristics, and performance claims for most standard off-the-shelf parts and raw materials.

• *Process capability listings*—including not only what processes are available to the manufacturer in-house but also what tolerances and other specs can be held by the specific equipment on hand.

• *Test simulation software*—including stress analysis, thermal tolerances, aerodynamic properties, and a host of other specialized applications.

• *Electronic development fixtures*—computer simulation software that performs the same functions as a physical mock-up, such as identifying physical interference points.

The Materials Used in Production

The materials used for actual production in the mid-1990s fall into two broad categories: the data and software used to manage shop floor operations and the actual raw materials converted into products.

Data and Software: Raw Materials of Production Management

First of all, the data and software used by mainstream American manufacturers to manage shop floor operations include the following:

- *Schedule data*—delivery schedules as well as individual production schedules are required. In MRP II (manufacturing resource planning) environments where a shop floor control module is used, schedules are maintained and generally utilized at the individual work center or machine level. This may also be true in less automated environments where these schedules are maintained manually in paper-based systems, often on schedule boards.
- *Inventory data*—particularly for raw materials and work in process (WIP). Storage locations, quantities, and stages of completion are typically required. Inventory audit trails of all issues, picks, disbursals, and consumption in next assemblies are also usually required.
- *Bills of material*—including quantities of each component required per assembly or batch, general parent-child relationships (as they pertain to assemblies and

The materials used in production.

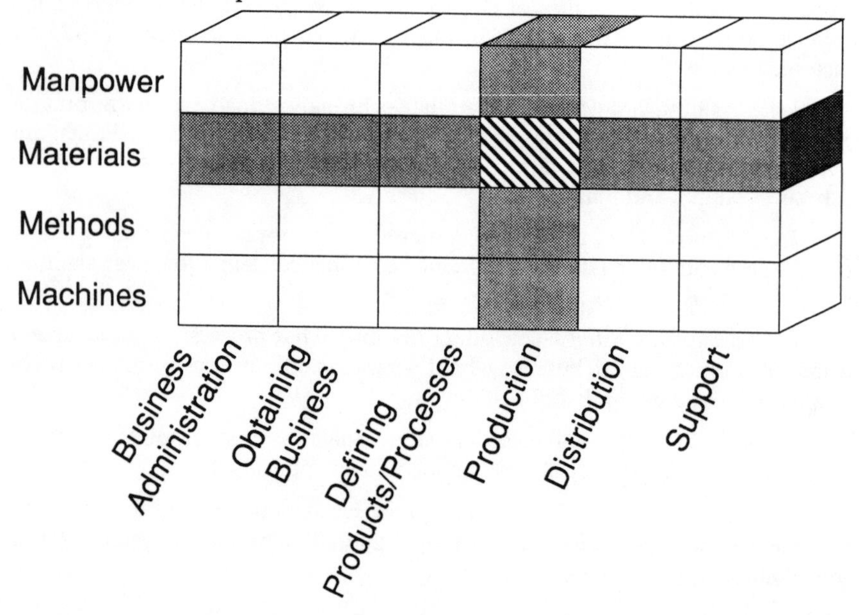

their components), and effectivities of specific components within designated assemblies.

• *Routing information*—for all parts and assemblies. This data defines which machines or processes are required to produce each part and assembly, in what order. Put simply, it tells the producer where to send each part from operation to operation, until it is completed and sent to shipping.

• *Labor and time standards*—for each manufacturing operation. This data lists the amount of labor time expected to be expended in the production of a part or batch for each manufacturing operation or process. It is typically expressed in minutes per piece, hours per hundred, and so forth. The data is typically developed and maintained by industrial engineers.

• *Labor accounting data*—including attendance and labor performance to the established standards just described.

• *Manufacturing methods data*—including assembly and installation drawings and general shop floor work instructions.

• *Quality assurance data*—including dimensional call-outs, tolerances, and other technical data.

• *Procurement data*—particularly related to vendor performance. Typically, this includes such information as delivery performance, purchased part quality history, price performance, and responsiveness.

• *Training materials*—especially those materials related to the training of factory mechanics, electricians, technicians, assemblers, supervisors, and inspectors. Over the last five years, with the drop in costs associated with video equipment, much of the training associated with these disciplines has moved toward video and away from the personal instructor and paper-based media.

• *Maintenance data*—including historical data for individual pieces of equipment, maintenance interval recommendations for each type of equipment, individual machine specifications, and preventive maintenance schedules.

• *Cost data*—generally a roll-up through the bills of materials, including material costs, labor costs, and some kind of overhead allocation by unit or batch produced.

The *Real* Stuff

After reviewing the data and software used to manage shop floor operations, we must consider the physical materials found in some condition as natural resources here on Earth that manufacturers convert into actual products (see Exhibit 6-2). As Forester (*The Materials Revolution*) stated, the science dealing with these materials is widely believed to be one of the three "megatechnologies" that will dominate the 1990s.

The classification of materials may be done in a number of ways. For our purposes here, eight broad categories are considered:

1. *Metals and alloys*—including steels, superalloys, and light alloys. Their principal characteristics are strength and toughness, although they are often

Exhibit 6-2. Significant materials characteristics required by American manufacturing industries.

	Aerospace	Automotive	Chemical	Electronics
High Strength - to Weight Ratio	X	X		
Temperature Resistance	X		X	
Corrosion Resistance	X	X	X	
Near - Net - Shape Formability	X	X	X	X

accompanied by significant weight penalties. Primary uses are in automobiles and aircraft.

2. *Plastics*—including polymers, rubbers, and polyurethanes. Their principal characteristics are strength, resistance to corrosion (especially oxidation), and low density. Primary uses are in automotive, electrical/electronic, coating, and construction industries.

3. *Ceramics*—including alumina, silicon nitride, and metal carbides. Their principal characteristics are hardness, temperature resistance, and corrosion resistance. Primary uses are in furnace refractories, cutting tools, and (increasingly) engine components.

4. *Composites*—including fiber-reinforced (usually glass or carbon fibers) plastics, metals, or ceramics. Their principal characteristics are toughness, strength, and low weight. Primary uses are in aircraft, automotive, and electronics companies.

5. *Paper*—including paper and paperboard. Their principal characteristics are low cost and application diversity. Primary uses are for printing and decorating.

6. *Fiber*—including cotton, nylon, and glass. Their principal characteristics are diversity of application and ease of handling. Primary uses are in textiles and plastic/fiber composites as described in item 4.

7. *Wood*—including wood and wood composites. Their principal characteristics are ease of fabrication and strength. Primary uses are for furniture and construction.

8. *Stone*—including building stone and cement. Their principal characteristics are durability, weight, low cost, and wide availability. Primary uses are in construction and transportation (roads, bridges).

To more thoroughly understand the specific relationships of production materials to the major industries using them, let's consider several of the industries in more detail.

• *Aerospace,* which employed about 835,000 workers in 1987 and had sales of about $105.6 billion. This industry selects materials primarily according to life cycle cost, strength-to-density ratios, fatigue life, fracture toughness, stiffness and strength, and low thermal expansion levels. Currently using mostly metals, alloys, ceramics, and polymer composites.

• *Automotive,* which directly employed about 1 million workers and had sales of about $223 billion in 1987. This industry selects materials primarily on the basis of raw material expense, ease of processing, material life, and reliability. Currently using mostly metals, polymers, ceramics, and glass.

• *Biomaterials,* which has estimated annual revenues of $50 billion (no employment estimate). The industry produces such items as artificial organs, medical diagnostic devices, cardiovascular and blood products, drug delivery systems, materials for plastic surgery, and prostheses. It also encompasses the field of biotechnology. In the more specialized applications, high costs are tolerated taking account of physical and biological properties. In consumables areas such as disposable hospital supplies, cost is more of a driver of material selection. Currently, materials used are composed largely of synthetic polymers, water-soluble polymers, biopolymers, metals, ceramics, glass, and carbon.

• *Chemicals,* which had sales of over $195 billion in 1987 and employed more than one million people. This industry selects materials primarily according to strength, corrosion resistance, and catalytic properties. Currently using mostly polymers, ceramics, composites, and single crystals.

• *Electronics,* which contributed about $155 billion to the U.S. economy in 1987 and employed more than 1.3 million workers. This industry selects highly engineered materials primarily on the basis of physical properties such as wafer flatness and larger wafer diameter, defect density, and the ability to produce increasingly small features. Currently using mostly silicon, gallium arsenide, and indium phosphide. Also uses many process chemicals such as solvents, acids, photoresists, film etchants, and gases for film deposition.

• *Telecommunications,* which employed one million-plus workers in 1987 and yielded sales of $16.1 billion in equipment as well as $130 billion in services. This industry selects primarily electronic and optical materials on the basis of quality. Most current telecommunications devices employ high-quality, single-crystal silicon. Quartz is also an extremely important material to this industry, and synthetic quartz has allowed the United States to become independent of overseas suppliers in this area. Currently using silica optical fibers, III-V semiconductors with indium phosphide substrates and gallium-indium-arsenic-phosphorus epitaxial layers.

The Incredible Shrinking Product

A general trend that has been widely followed over the last two decades is described as "dematerialization." Examples of this phenomenon abound, but the most common is probably the American automobile. The 1976 Chevrolet Caprice Classic, for instance, has a curb weight of 4,424 pounds. The 1986 model of the same car has a curb weight of 3,564 pounds, a nearly 20 percent reduction. The bulk of the weight reductions in cars were attributable to decreases in use of steel and iron and increases in the use of aluminum and plastic.

Ward's Automotive Yearbook in 1988 indicated that the total pounds of material used in the production of a "typical U.S. car" was as follows:

1978:	3,569.5
1984:	3,232.0
1986:	3,170.5
1988:	3,167.0

This reflects an average decrease per automobile of about 11 percent.

While the experiences of the automotive industry over the last two decades may be attributed at least in part to the energy shortage that drove much of the economy in the 1970s and 1980s, dematerialization is a much broader and more pervasive trend, and it is certainly not restricted to the automotive industry. In fact, there are two fundamental reasons why this phenomenon has occurred and continues to play such an important role in defining the products and processes of U.S. manufacturers:

1. U.S. (and to some varying degrees, foreign) consumption of goods increasingly consists of goods with high levels of value-added content. More technology, more information, and less purely physical material comprise these products.
2. Products and services are simply allowing more efficiency in performing their target functions. (A clear example is the roughly 98 percent reduction in weight requirements achieved in telecommunications through replacing copper wire with optic fiber.)

This trend has manifest itself clearly over the last two decades, and as we will see in subsequent chapters, is likely to continue into the foreseeable future.

A major factor contributing to dematerialization is the dramatic improvement over the last several decades in strength-to-density ratios of the materials used. The ratio of material strength in our most advanced structural materials (aramid fibers and carbon fibers) is now almost 50 times greater than that of steel in the year 1800, and many times that of aluminum in the year 1900.

As an aside, it is also interesting to note the levels of import dependency of American manufacturers, especially relative to our Japanese and European Economic Community (EEC) competitors. For example, for metals the dependency upon imports of the United States, Japan, and the EEC is as follows:

	U.S.	*Japan*	*EEC*
Aluminum	85%	100%	61%
Copper	14%	94%	90%
Iron ore	36%	99%	79%
Tin	83%	98%	87%
Zinc	57%	74%	91%

It seems reasonable to believe that, as the exporting countries endeavor to improve their economic positions, they are likely to demand that higher percentages of these materials be imported at a stage further along in the value chain (i.e., preprocessed and more costly forms of the same materials).

The Materials Used in Distribution

Materials used in the distribution activities of U.S. manufacturers through the 1990s include data and packaging materials.

Data

Among the data used are these:

- *Delivery schedules*—including both complete finished goods deliveries and service parts.
- *Market and sales forecasts*—particularly by geographical region; used for forecasting manpower and other capacity/resource requirements.
- *Route data*—defining shortest and most cost-effective distribution routes for all proven and potential markets (generally paper maps).
- *Cost data*—historical actual costs by distribution medium, carrier, and area. Also current cost listings by carrier, considering weights, premiums, and other relevant factors to derive a comparative extended cost index.
- *Third-party data*—including historical and current data pertaining to all aspects of distribution of not only the manufacturer's product but also, at times, like products of all manufacturers distributed through the same third party.

The materials used in distribution.

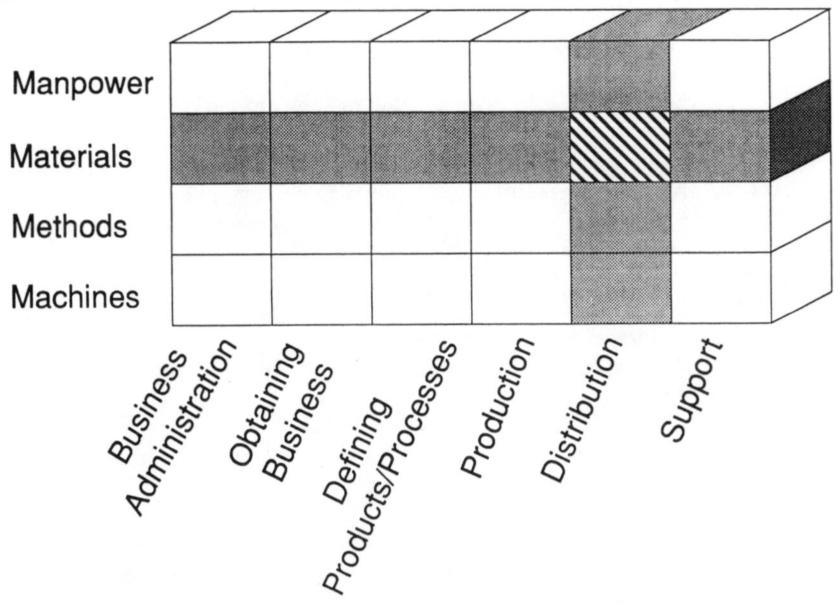

• *Field performance data*—focusing on failure levels of product and part types (particularly by geographical area where this is relevant).

• *Inventory data*—which describes where inventory is located, how often inventory is turning over, and where inventory level is disparate with demand.

• *Delivery performance data*—generally reflected as percentage of shipments delivered by scheduled delivery date, broken down by region or carrier, and so forth.

• *Bar code or other parts ID schema*—the method and applicable standards used to code the parts for rapid identification during receiving, storing, picking, and issuing parts/products for delivery and for inventory-auditing activities.

• *Inventory management software*—for analysis and appropriate distribution/redistribution of finished goods and service parts inventory. Also serves as a feed to accounting systems for inventory-costing purposes.

• *Order tracking and management software*—to manage customer orders and ensure the earliest possible delivery of complete orders.

• *Product design and packaging information*—which calls out such data as the physical dimensions, weight, shelf life, and packaging requirements of each product or part.

Packaging for the Green Consumer

In addition to the data required, there is a substantial amount of packaging material used in the distribution of most products. In recent years, there was a marked shift in many cases from purely paper-based packaging materials toward plastics and foams. Styrofoam, cellophane, and plastic tie straps are among the most popular. However, as the U.S. consumer becomes more environmentally aware, the levels of plastics in packaging is ebbing in some areas. (Someone recently described the dilemma in terms of "whether to kill rain forests or clog land fills," referring to the choice between paper and plastic grocery bags.)

Packaging materials fall into several categories, including:

- *Containers*—e.g., corrugated boxes, mailing tubes, plastic containers, wooden shipping crates, tote boxes, metal containers, pallet bases
- *Cushioning materials*—e.g., cushioned pouches, mailing bags, cellulose wadding, air bubble wrap, polyethylene foam, flowable cushioning, molded cushions
- *Barrier materials*—e.g., barrier bags, barrier films, poly bags, tubes, tarps, desiccants, storage drums, corrosion inhibitors
- *Magnetic shielding materials*—RFI (Radio Frequency Interference) shield, EMI (Electro Magnetic Interference) shield, magnetic shield, electrostatic discharge shield, and radiation shielding materials
- *Packing materials*—tapes, labels, adhesives, shrink-wrap, protective netting, banding (metal and plastic), and wadding

The major changes in packaging materials, then, have been in strength, rigidity, and weight reduction. As we move through the last half of the 1990s, we are likely to see shifting emphases toward recyclability and improved shielding from RFI and EMI.

The Materials Used in Customer Support

The materials used in customer support activities generally consist of data regarding product specifications, service-related training materials, technical publications, and service-related software. Among them are the following:

- *Service part sales data*—including part types, sales volume, region, product failure cause data, and percentage of sales covered by product warranty.
- *Repair response time data*—historical performance data related to how long it took to respond to service/repair requests
- *Repair parts*—both service parts commonly defined as such and stocked for that reason and "special circumstance" items where a customer requests some subset of the product not normally stocked as a replaceable item
- *Repair equipment records*—including inventory of all repair equipment, cross-references to what products can be repaired with what equipment, and preventive maintenance schedules/directions for each piece of repair equipment
- *Customer service crew management information*—an index of which personnel are qualified to repair which products, each worker's certifications, training requirements, and so forth
- *Crew management software*—to plan, track, and manage the deployment of service crews, as well as to track the costs associated with service activities

The materials used in product and customer support.

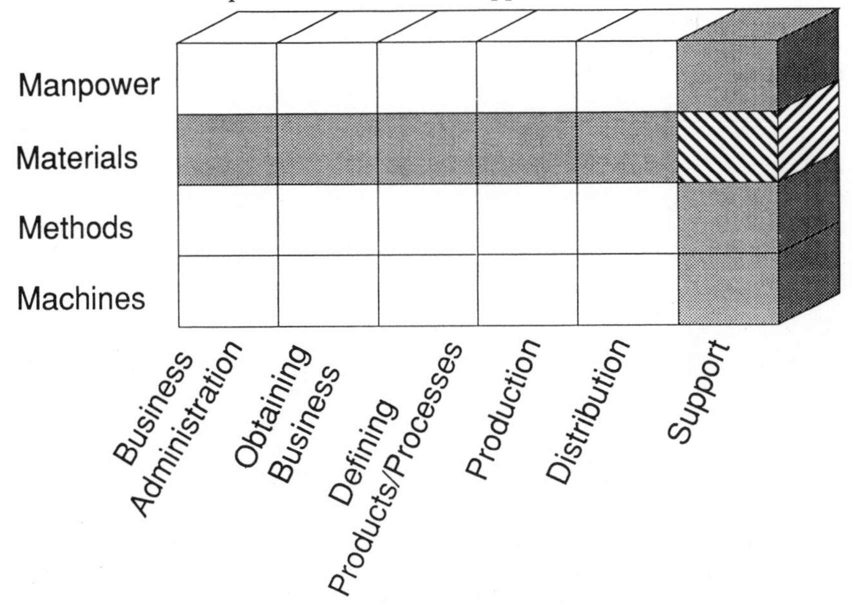

- *Service agreement and warranty management data*—listings of what services are covered by what types of service agreements and warranties, the effective life of each agreement or warranty, and what work has been performed against each type
- *Technical publications, repair manuals, and other service instructions and documentation*—as mentioned earlier, these items have become video-based in many cases over the last five to seven years
- *Training materials*—videos, books, and class materials required to support the training and management of service technicians
- *Performance management software*—software that tracks customer service performance in terms of cost, speed, quality, and customer satisfaction

In recent years, the lion's share of attention and improvement activity within the customer and product support realm have concentrated on the product and the customer. Sounds logical, doesn't it? We'll see in subsequent chapters why that won't be enough to succeed in the future. The degree to which we are successful in this area over the next decade will more often be linked to how well we manage something called the "infrastructure." All of our distribution and service expertise will still be important and relevant, but it won't be enough without this other catalytic element.

Chapter 7

Dominant Machines: The Reign of the Computer

Notwithstanding all that's been said up to this point about the importance and the impact of the computer in American manufacturing over recent decades, it must be restated again here: Nothing has had a more powerful, more sweeping, more important impact on American manufacturers (and American business in general) than the computer. There are four primary reasons for this.

First is the obvious benefit of sheer number-crunching speed. The volume of transactions now processed by computers is measured in MIPS (millions of instructions per second). Second, the "user friendliness" of modern computers is astounding. With only rudimentary training, any ordinary individual today can learn to use popular software packages such as spreadsheet programs and word processing programs in a matter of hours. Third, the accessibility of computers is astounding. Virtually every workplace and many homes now have them. In fact, for PCs (personal computers) alone, the number of units in use soared from about 3 million in 1981 to over 57 million in 1991. Fourth, the applications for computers and computer technology are almost limitless. Looking at Exhibit 7–1, the reader can quickly see that computer technology appears in major roles in every phase of the manufacturing business. It is used by everyone from the CEO to the shipping clerk. There is no more versatile machine.

There have, of course, been great strides in other areas of manufacturing machine technology as well. Not only has the computer been incorporated into machines like milling equipment and computer-aided design workstations, advanced materials have been incorporated into these pieces of equipment as well. High-strength metal alloys, composite materials, ceramics, optic fibers, and various new polymers have been incorporated to reduce the size, weight, and cost. Beyond the impact of materials and computer technologies, the *processes* of manufacturing are also in the midst of dramatic change. This means the equipment used is a hybrid of old machine technologies and new technology specifically tailored to perform or support the new processes.

Still, with all of the progress that has been made, there is far more to come. But that is the subject of Chapter 14. For now, let's consider the state of machines as they exist today in the American manufacturing environment (see Exhibit 7–1).

The Machines Used in Business Administration

The machines currently used in business administration activities in American manufacturing are primarily office equipment and what has traditionally been

Exhibit 7-1. Dominant U.S. manufacturing machines.

Business Administration	Obtaining Business	Defining Products and Processes	Production	Distribution	Support
Computers - Mainframes - Minicomputers - P.C.s Local Area Networks Modems Video Conferencing Equipment Faxes Copy Machines Printers Word Processors Typewriters Scanners Telephones, Mobile Phones, PBXs Optical Disk Drives Pagers Overhead Projectors	On-Line News Services On-Line Demographics Pagers Point - of - Sale data Collection Devices Videotaping and Editing Equipment Brochure Printing Equipment Computers - Mainframe Hookups - P.C.s - Printers - Modems Local Area Networks Videoconferencing Equipment Faxes Word Processors, Typewriters	Computers - Minicomputers - P.C.s - Work Stations - CAD Terminals Local Area Networks Printers / Plotters Copy Machines Faxes Optical Disk Storage Devices and drives Test Equipment Digital Translators Blueprint Machines Aperture Card Readers / Printers	Computers - Minicomputers - P.C.s Local Area Networks Blueprint Machines Aperture Card Readers Fabrication Equipment Assembly Equipment Inspection Equipment Facilities Labor Collection Terminals Inventory Management and Production ReportingTerminals Material Handling and Transportation Equipment Telephones, Pagers and Faxes Bar Code Readers	Computers - Minicomputers - P.C.s Telephones, Pagers and Modems Aperture Card Readers, Blueprint Machines and CAD Terminals Bar Code Readers and Remote Scanners Automated Storage and Retrieval Systems Material Handling and Transport Equipment Trucks, Trains, Planes and Boats	Computers - Minicomputers - P.C.s - Printers Telephones, Pagers and Modems Aperture card Readers and Microfiche Readers Diagnostic Equipment Installation and repair Equipment and Tools Technical Publication Printing Equipment Video Recording and Playback Equipment Service Fleet Vehicles

The machines used in business administration.

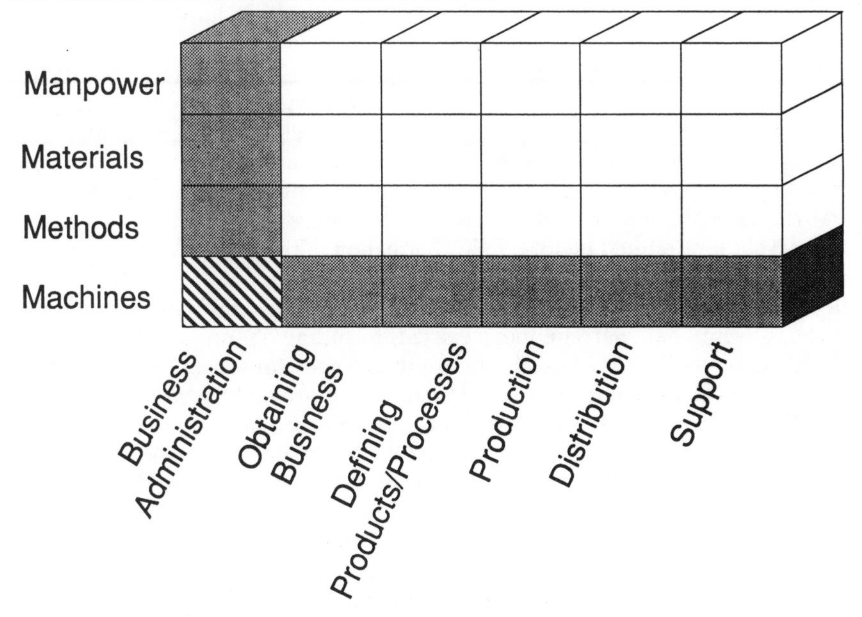

described as data processing equipment. There is, of course, a good deal of overlap between these categories, owing mainly to the advances in computer technology over the last two decades.

Negotiating the Data Processing Maze

Data processing up until about 1980 was a subject that evoked images of some artificially cool, clean, isolated set of offices on elevated floors with heavy wires spiderwebbed beneath them. The offices were occupied by large, intimidating metal boxes that whirred and belched out reams of green and white paper. The people who worked in these shrines to electronic magic were regarded as iconoclasts, people who chose to leave the real world behind. These folks spoke their own special language, and only another DP person could understand them. They thought and spoke in terms of JCL (job control language), ROM (read only memory), RAM (random access memory), bits (*bi*nary digi*ts*), bytes (eight-bit words), and nibbles (half a byte, of course).

The 1980s brought about changes in computer hardware (and just as important, in computer software) that made the devices both inexpensive enough and user friendly enough for almost everyone. The result has been what is referred to by Alan Hald of Microage, Inc., as EUC (end user computing). People no longer have to speak the computer's language in order to effectively use one.

To a large extent, computers now speak English (or Japanese, or what have you). This, in combination with the tremendous advances in miniaturization of the circuitry involved, has served to blur the once-distinct lines between what we used to refer to as four distinct classes of computers: microcomputers (personal computers, or PCs), minicomputers, mainframes, and supercomputers.

The last time I looked seriously at the breadth of hardware and software applications available for manufacturing operations was in the mid-1980s. At that time, I found that there were about seven different viable microcomputer-based systems available that could adequately support a small manufacturing operation, including units produced by IBM, AT&T, Altos, NCR, and Sperry. Prices for these units ranged from $7,000 to $100,000, and memory ranged from 1 megabyte to 16 megabytes of RAM. As for minicomputers, there were at that time about twenty potential solutions offered by IBM, DEC, Formation, Nixdorf, HP, Wang, and MAI. These devices ranged from $16,000 to more than $1 million, with available memory of between 1 megabyte and 64 megabytes of RAM. In the mainframe area, I found that about twenty hardware solutions existed, including offerings by Sperry, DPS, IBM, HP, Quantel, DEC, Amdahl, and Univac. Prices for the units ranged from just over $100,000 to well over $13 million, and memory ranged from 2 megabytes to 128 megabytes of RAM. The average processing capability (as nearly as I could determine) at that time was between 2.5 and 3.7 MIPS (millions of instructions per second). Some of the best computer systems by 1990 were able to perform at more than 50 MIPS. Some folks at the World Future Society's FUTUREVIEW in July 1989 projected that by the year 2000, computers will be running at about 2,000 MIPS.

It may be useful to compare cost per MIPS over time. In 1960, the relative cost of computing on a mainframe in $/MIP was $10 million. By 1980, it was about $600,000. By 1990, it had dropped to around $100,000.

A few words about supercomputers: Existing supercomputer technology is neither practical nor affordable for manufacturing operations. However, as a benchmark, supercomputer technology as it existed in the mid-1980s approached 100 MIPS. These devices are generally used primarily by government agencies and universities.

The primary technology responsible for these dramatic improvements has been semiconductor technology. Faster chips are increasingly required to support more and more imagery in computing, simulations and complex modeling, and graphic presentation of scientific analyses. Semiconductors not only provide these faster speeds, they have also enabled the miniaturization of chips.

This miniaturization, in turn, allows for more and more chips to be packed into smaller and smaller devices. It has also allowed for more memory to be placed in smaller physical areas, by stacking multiple layers of integrated circuits onto single chips. As a benchmark of where the industry stands today, Hitachi announced in June 1991 that it had completed the first working prototype of a 64-megabit chip (64 million bits of information).

In mainstream American manufacturing environments, it is common to find multiple microcomputers, "dumb terminals," printers, and other input/output devices connected by a LAN (local area network). The LAN supports common use of devices (for example, it often allows several different microcomputers to utilize the same printer) and data files. It can also serve as a "gateway" to the large minicomputer or mainframe computer systems, feeding information into and out of the LAN (known as "uploading" to and "downloading" from the parent device).

It is common to find that purveyors of CIM (computer integrated manufacturing) offer up complex, tiered architectural schemes for computerization that depict

PCs at the bottom, minicomputers in the middle, and mainframes at the top. All of these devices are interconnected in parent-child relationships that appear to be perfectly rational and logical. However, reality in most manufacturing environments is usually somewhat different. It is quite common to find stand-alone systems, user-modified systems, and general data flow breakdowns throughout the networks, with compatibility problems experienced in both hardware and software.

There are several reasons for this situation. Over the last decade, a wealth of new software has been developed that is tailored specifically to various manufacturing and manufacturing support applications. Of course, they all run on different operating systems, which means different physical hardware. When an end user identifies a software package that is the "perfect solution" for his or her particular need, the odds are seldom better than 50/50 that it will be consistent with the long-range systems architecture strategy of the information technology department. Hence, a constant battle ensues, and systems fragmentation occurs.

Therefore, at this juncture, there are few really well designed, fully integrated companywide systems strategies. The only real solution to this problem will be standardization, allowing devices with different operating systems to communicate and efficiently share information.

New Tricks for Old Dogs

In terms of office equipment, the mainstream American manufacturer has a plethora of equipment at his or her disposal. The equipment that has proved most useful is tailored to meet what have been long-standing needs in support of the administration of general business.

For example, the word processor has now almost made the traditional typewriter obsolete. Typing has always needed to be done. What the word processor has contributed, however, is the capability to easily edit and spell-check in-process work. In addition, word processors allow for multiple fonts and the easy creation and editing of various business forms.

Another of the most common and most vital devices to American manufacturing businesses is the telephone. No longer is this a simple instrument with a rotary dial, however. Your local telephone company representative (now often a service center retailer) would be happy to walk you through a wide range of device alternatives, including speaker phones, call waiting options, call forwarding, caller ID, automatic redial, backlit touch-tone buttons, memory-based speed calling, volume adjustments for the hearing-impaired, multiple lines, mute functions for private off-line conversations while the call is in progress, hold buttons, and combination phones incorporating clocks, tape recorder answering machines, and AM/FM stereos. Even more features, such as electronic voice messaging and automatic message distribution, are available through PBXs (private branch exchanges). And as if all of this functionality weren't enough, business magnates now never have to be without them. Cellular phones and other mobile telephone technologies have ensured that we never have to be out of touch or unreachable. Even if you're not exactly a magnate, just about all businesses these days can afford a pager or two. (As my grandmother often said, there is no rest for the wicked.)

Then there is the fax (or data-fax or facsimile) machine. When the communication needs to be textual or graphic (or some combination thereof) and needs to get somewhere fast, there is simply no beating the fax. I first really came to grips with the importance and proliferation of fax technology when I saw my fax number was automatically added to my business cards by my employer in 1990. Advertisements are even being faxed now, giving rise to a new form of "junk mail"—junk faxes! Especially in dealing with foreign suppliers, customers, and business associates, fax machines are becoming an indispensable tool for supporting ongoing business operations.

Probably even more commonly used than the telephone these days is the copy machine. Thus far in the 1990s, paper is still the medium of choice for most people when it comes to data transfer, record creation, and conveying masses of information between computers and people (e.g., exception reports, sales figures, and so on). As long as this is the case, there will be a tremendous need for copy machines. Copy machine technology has also come a long way from its humble beginnings. These days, copiers can handle multiple colors, generate transparencies for overhead projectors, feed devices that generate color 35mm slides, enlarge, reduce, darken, lighten, collate, and staple. They perform self-diagnostics and walk their maintainers through the required remedies to paper jams and other problems step by step with both textual and graphic aids. They are many times quieter and cleaner to operate than they were several years ago and are generally more compact. They are also far less costly on a function-to-cost basis.

Probably the most common medium for displaying data (both textual and graphic) to several people at a time is the overhead projector. Production of the transparent slides is relatively inexpensive and can be done by just about anyone who can run a photocopy machine. The slides can be made in black and white or color and are almost as easy to carry and use as paper. For flashier productions where the same presentation is likely to be made over and over again, 35mm slides are often still the medium of choice. Although the projection equipment is more complex and expensive and is generally more prone to jams, scores of slides can be transported in a single tray. In addition, the color and resolution of 35mm slides is much more striking and is far more suitable to presentation of complex photographic images. The 35mm slide is generally more time-consuming and expensive to produce than the overhead transparency.

When it comes to training aids, or when the most powerful nonlive presentation is required, or when the objective is to make a relatively "personal" statement, there is simply no beating full-motion video with sound. Today, it is rare to find a business that doesn't have on-site video display equipment, generally in the form of a VCR (videocassette recorder) and a portable color television. Some major corporations are now sending out their quarterly financial reports on VHS videocassettes. Many are using video-based training for everything from presentation skills classes to final assembly operator training. Still others use the videocassette as a marketing device.

A more recent technology related to video-based training is the live, multisite videoconference. The videoconference allows people in different locations (literally, even in different corners of the world) to communicate live via actual video images and voice teleconferencing. While expensive by normal telephone standards, it often yields impressive ROI when contrasted with airline tickets and lodging for participants' travel to one another's location for personal meetings. As the supplier base and customer base of major manufacturing concerns takes on an increasingly global complexion, videoconferencing becomes increasingly attractive. The up-front investment to maintain on-site videoconferencing facilities is still quite substantial, however, and this has slowed the growth of the medium significantly.

A Look Into the Executive Toy Box

For the ultimate in office equipment, we come to the category of executive toys. Although it is hard to imagine life in the office (or anywhere else, for that matter) without the pocket calculator, we have reached a level of indulgence in America where calculators are built into wristwatches, notebook covers, rulers, and business card holders. There are now occupying the briefcases and desk drawers of managers everywhere "handy" little devices that display the time in any time zone in the world, keep track of calendars, type out short memos, electronically store business card data, and even keep track of telephone numbers that might formerly have been the domain of a "little black book." Some devices truly are helpful. Others are more like the handy kitchen appliances so often demonstrated on TV around Christmas time. The ones that slice, dice, and do julienne fries. The ones so often found in garage sales the following June.

The Machines Used in Obtaining Business

The machines involved in the "obtaining business" portion of manufacturing fall into three basic categories: computation, communication, and publication. Each of

The machines used in obtaining business.

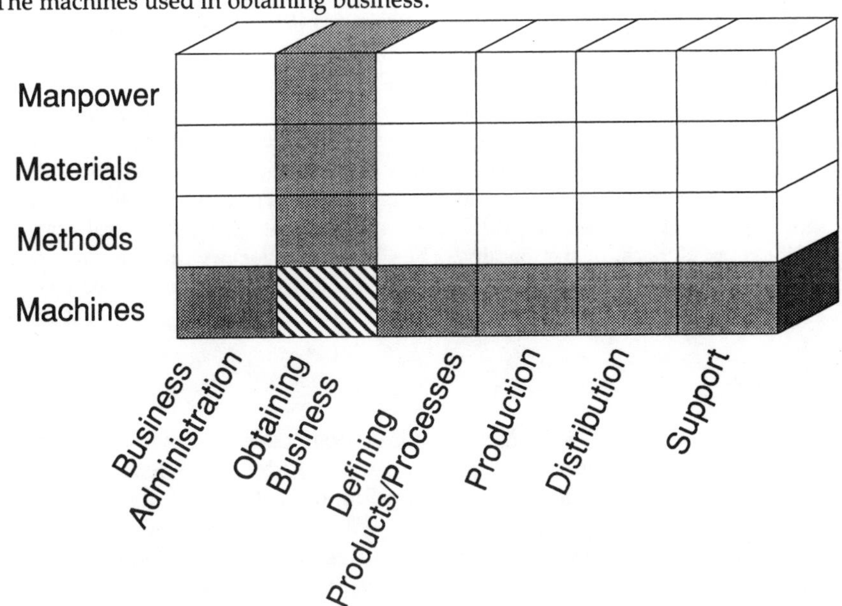

these areas, as with all of the "machine" categories in Exhibit 7-1, have been greatly affected by computer technology.

The Forecast for Forecasting

In the category of computation, the most physically demanding functions from a machine standpoint are generally those involved in sales forecasting and market analysis. The methods used vary, of course, depending on the nature of the manufactured product. Generally, the closer the manufactured product is to the end of the overall value chain, the more detailed and intense this activity becomes. (A manufacturer of automobiles, for example, would be more heavily involved in market analysis activities than a manufacturer of automobile frames.)

Once sales figures are accumulated and loaded into a spreadsheet, most fundamental projections for sales can be performed by a good personal computer. Several software packages available today for PCs do a good job in this area and run on most of the popular operating platforms. Almost any standard PC with a hard drive can be used to perform the basic forecast computation techniques (e.g., exponential smoothing, regression analysis). More complex analyses such as econometric modeling and business cycle forecasting techniques require more computer "horsepower" than has been available in most PCs to date.

However, as we discussed previously, the power and capacity of computers is increasing so rapidly that the lines between types of computers have blurred. It is no longer possible to specify what class of computer will be required for these applications. It is only appropriate to state that even the most complex modeling techniques we use today will not be out of reach of end-user computing for very long.

Communicating the Sales Data Needed for Forecasting

Capturing and communicating the actual sales data from the field to the sales forecaster's computer is another extremely important aspect of the "machine" requirement for obtaining business. It falls into the category of communication machines/equipment. The complexity of this data accumulation effort depends on the levels of hierarchy in the sales end of the business. For example, a factory may need only to combine the individual forecasted requirements of several regional sales branches, depots, or dealerships. Then again, the factory may be required to do full-blown forecasting for national and even international sales on the basis of data collected from hundreds of thousands of individual point-of-sales locations.

Bar Code Literacy

Where point-of-sale locations are utilized, bar code readers, OCR (optical character recognition) devices and magnetic stripe readers have become essential pieces of equipment. Retail outlets quickly found in the 1970s that data entry errors, time expenditures, and training requirements were simply too daunting to allow automated data collection from point of sales to be efficient. Particularly with the advent of bar code scanners, most of those problems have been eradicated.

Generally, the point-of-sale information is held in a memory repository at that location for a daily, weekly, or monthly upload to regional or company data bases.

The uploading function is generally conducted over telephone lines (sometimes dedicated lines, but not usually) via EDI (electronic data interchange). EDI involves the electronic transfer of information from one point to another. This generally involves a computer and a modem at each end of a telephone line (similar to a fax machine).

The combination of EDI and bar coding present tremendous potential benefits not only in the area of gathering and communicating data for sales forecasting but also in the areas of inventory management and retail merchandising labor. An excellent book by Norman Weizer entitled *The Arthur D. Little Forecast on Information Technology & Productivity* (Wiley, 1991) describes the efforts of Levi Strauss & Co. in this area. Weizer makes the following observations about the company's use of this technology via its Levilink program:

- Levilink pretickets merchandise with bar codes, enabling retailers to display the clothing without tagging it again in their stores. (This process saves retailers three to fourteen days in the time it takes to display products.)
- Levilink generates electronic packing slips that arrive before a shipment, listing all merchandise in advance of its receipt. (The packing slips also function as electronic invoices, reducing paper handling.)
- Levilink automatically tracks sales to reduce the cycle time on reorders.

The technology utilized by Levilink in these processes includes bar code scanning, EDI, point-of-sale data collection, and electronic funds transfer.

Other information required by sales forecasters includes market data, demographic data, and general news about their specific industries. Often, this information is obtained through on-line news services like the Dow Jones News Service, Compuserve, and Prodigy. Again, to obtain and use this data, a modem and a PC are generally all that is required.

Where Computation Meets Communication

An interesting and promising hybrid of the categories of computation equipment and communication equipment in the area of obtaining business is the field-deployed order entry device. Particularly well-suited to the new notebook-style computer, these devices configure, enter, record, and communicate orders. They not only speed input but also allow visibility of "available to promise" inventory, current order-through-delivery lead times, and current prices. They also prevent misconfigurations, incorporate latest design revisions, and so forth for presentation to potential customers.

The balance of the communication-related equipment required in the process of obtaining business (for example, telephones, faxes, pagers) is common to the rest of the organization and is dealt with in the section on the machines used in business administration earlier in this chapter.

Have Notebook, Will Travel

The third and final category of machines required for obtaining business is the equipment involved in various aspects of publication. Sales brochures, new product announcements, and other marketing publications require textual, graphic, photographic, and video and sound production and presentation equipment.

A real boon in this area over the last five years has been the development of fairly friendly desktop publishing software and laser printing technology. Never has the production and merging of text and graphics been so easy and so accessible as with the simple combination of a PC, desktop publishing software, and a laser printer. The quality of the final product looks as though it was produced by a professional printer, and the cost of these elements is now within the range of even the smallest manufacturers.

In terms of personal sales tools, new notebook PCs now have the capability to display full-motion video and stereo sound. Salespeople can demonstrate a new product in full color video on the PC screen, with narrated voice-over, freeze-frame, and fast-forward or reverse. The entire package, including voice and video, can be stored on a standard 3.5-inch diskette. In addition, there is now a peripheral device available that allows the computer display to be interfaced with a standard overhead projector, so that a room full of people can watch the display simultaneously.

The Machines Used in Defining Products and Processes

The equipment currently used by mainstream American manufacturers for product and process definition includes several different platforms for electronic product definition and electronic process definition.

Letting the "Mouse" Build the Mousetrap

Electronic product definition machines are almost all derived from the apparatus described as computer-aided design (CAD) equipment. CAD was developed by very large companies such as General Motors and Boeing in the late 1950s and 1960s, utilizing large (by the standards of that time) mainframe computers. Because the development of CAD was very expensive (owing to computing hardware and software programming costs), most of the development work was sponsored by defense contractors during that period. CAD really emerged as a marketable product in the 1970s, as costs associated with the demanded computer power began to come down. Between 1973 and 1983, the CAD market jumped from about $25 million to over $1.6 billion. Later in the 1970s, and throughout the 1980s, industry-specific CAD systems were developed offering both two-dimensional and three-dimensional capabilities.

Reduced to its basics, CAD is an electronic drawing board for product designers. Engineers use a "mouse," the keyboard, or an electronic "light pen" to direct computers in combining lines, curves, and dimensions into product designs. More recent and sophisticated systems display the electronic "drawing" in three dimensions, and even rotate it in all three dimensions on the screen as the

The machines used in defining products and processes.

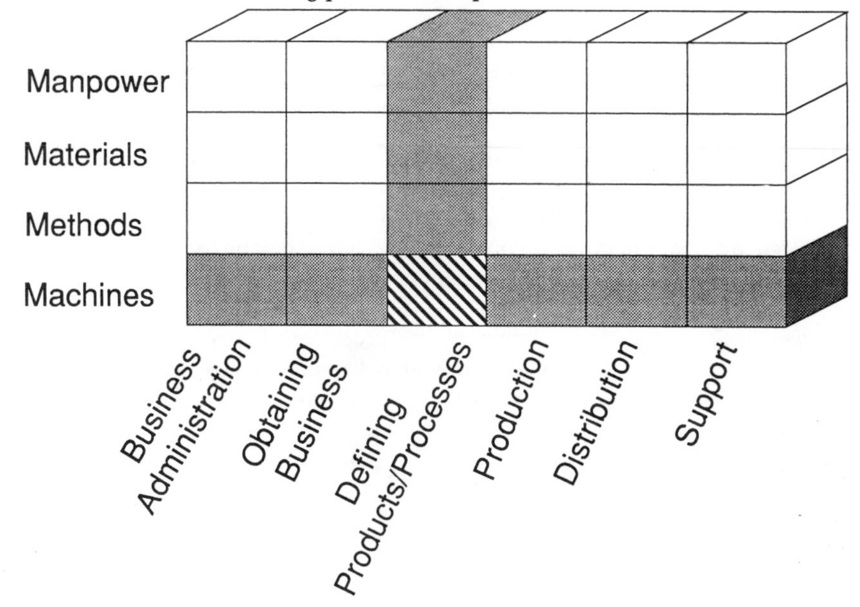

engineer watches. These machines can shade the drawings to appear as they would physically be seen by the human eye. They can enlarge, reduce, and "paint" surfaces on the products drawn, creating three-dimensional part models. In addition, the most current systems perform functions such as solids modeling, finite element modeling, logic simulation, and associative data base management.

Hardware continues to be the largest cost element for the bulk of these systems. The major vendors in that market include DEC, Sperry, Honeywell, Prime, Data General, Hewlett-Packard, Harris, and Perkin-Elmer. Alliances between users and software developers have sprung up over the last two decades and have produced some of the best CAD systems in use today. Among them are Prime Computer and Ford, and Control Data and Chrysler. Two of the best and most popular packages in aerospace and defense are CATIA (developed by Dassault of France) and Unigraphics (developed by McDonnell Douglas). Another aerospace and defense system is NCAD, developed by Northrop. Depending on the system size and features, they can range in price from under $20,000 for a microcomputer-based system to well into six figures for mainframe-based systems.

The output from these devices can be generated in the form of paper prints or electronic drawings and models. Paper prints are typically generated by plotters and can be produced in multiple colors. Microfilm and microfiche "prints" can also be produced by some systems, as can photographs. Electronic drawings and models can be retained on disk, improving portability and storage capability.

Processors for Process Design

For process development, computers are also the most fundamental design tool. The most advanced equipment for manufacturing in virtually every industry is computer controlled. Numerically controlled equipment cuts, shapes, and punches metal. Digital input devices initiate, monitor, regulate, and terminate the flow of ingredients in food and chemical industry equipment, as well as the temperature, humidity, and even pigmentation content based on color in film and paint manufacturing and processing. In metal production, furnace temperatures and other critical factors are computer-controlled as well. In assembly environments, a device called a programmable logic controller (PLC) allows assembly equipment to be driven by computer.

In all of these cases, the capabilities of the equipment involved can be monitored by computers, with process capabilities documented so that future designs can be produced with the limitations of existing equipment in mind.

Beyond simply documenting and accounting for process capability in new and modified designs, the development of the processes themselves often involves process simulation and modeling. These activities are usually performed on PCs and workstations otherwise utilized for CAD functions.

The Machines Used in Production

The machines used in production fall into two broad categories: shop floor management equipment and actual conversion equipment. As we move through the 1990s, the lines between these categories are beginning to blur, again largely owing to the incorporation of computer technology. However, the categories are still distinguishable at this point.

The machines used in production.

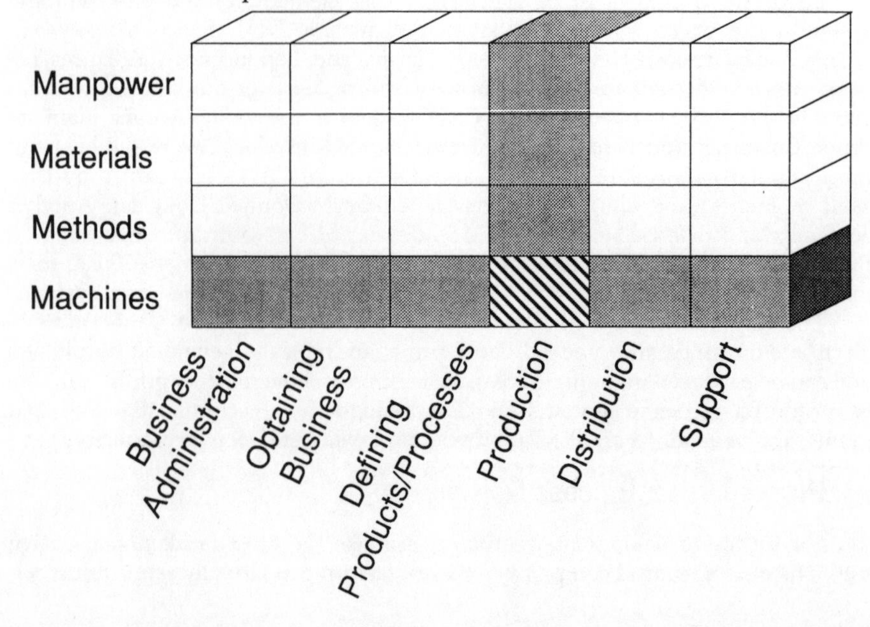

When LANs Man the Shop Floor

Shop floor management equipment typically includes computers (PCs), LANs, printers, plotters and blueprint reproduction machines, copiers, aperture card readers, labor collection devices, production reporting devices, telephones, bar code scanners and printers, and pagers.

PCs are often used on the shop floor to track labor utilization, material availability, quality performance, manufacturing cost trends, and general performance to schedule. Most of this information is generally uploaded by LAN to a minicomputer or mainframe computer, analyzed, and returned in some more structured and uniform format to shop floor management displays. Printers are used to produce hard copies of this data, as well as to generate "move tickets" to attach to loads of work-in-process for material handling.

Plotters and blueprint reproduction machines are used to provide hard copies of part and assembly drawings to direct labor employees and inspectors to support their daily work activities. Aperture card readers are used in those instances where it is necessary to review the overall shape of a part for identification purposes (such as at the receiving dock), or to look up a specific dimension for inspection.

Smart Uses for "Dumb" Terminals

Labor collection devices are generally used to record attendance and production performance data by specific employee. The device used is frequently a "dumb terminal" linked by a LAN to a minicomputer or mainframe computer. The most capable devices generally utilize the employees' badge for both security and speed, reading a magnetic strip ("mag stripe") or bar code on the badge to identify the employee.

Production-reporting devices are also sometimes "dumb terminals" possessing little or no information-processing "intelligence" themselves, but simply acting as an input/output device for a local minicomputer or not-so-local mainframe computer. Generally, the production reporting device works something like this:

1. Employee approaches terminal and inserts badge into badge reader, or wands bar code from badge.

2. Terminal recognizes employee and displays menu on screen from which employee may choose a transaction type (e.g., report in to work center, report out of work center, report work completed, request new work assignment).

3. Employee selects transaction type by moving the cursor until that transaction name is highlighted or by pressing on the appropriate transaction name, if the device is equipped with a touch-sensitive screen.

4. Transaction format appears on screen, and employee types in required information (e.g., number of pieces completed, number of pieces scrapped, number of individual lots or loads or containers).

5. The device transmits this information via LAN to the central processor (mini or mainframe) or, in some cases, processes the information itself. It then sends a message to the local printer, which generates a move ticket to attach to

each container of material for identification of the parts, their stage of completion, and their next destination.

6. Employee detaches the tickets from the printer and leaves to tag the material to be removed from the work center or manufacturing cell.

The balance of machine types (e.g., telephones and pagers) used for shop floor management are described in the section on the machines used in business administration earlier in this chapter.

Manufacturing Conversion Equipment

Actual manufacturing conversion equipment is even more diverse than the manufacturing methods employed. As discussed in Chapter 5 on "methods," there is a large number of diverse manufacturing processes employed in mainstream American manufacturing. Although many of the most common are listed in Chapter 5, it is not our purpose to capture them all here. However, for those that are listed, it would be useful to get a sense of what types of equipment are required:

Metal Conversion Equipment

• In sheet metal processing, hydraulic presses are by far the most common machines used for shearing, punching, stamping, forming, bending, and hammering metal into a desired shape. Other equipment in the metal-forming field includes spin lathes for metal spinning and high-temperature presses for superplastic forming. Inspection equipment frequently employed to check the products of these processes includes standard mechanical inspection equipment—e.g., CMMs (coordinate measuring machines), micrometers, calipers, and scales—magnetic particle detectors, templates, and tensile and peel test equipment.

• In metal casting, typical equipment includes continuous casting equipment, melt furnaces, core casting machinery, cope and drag equipment, foam molds, slurry mixers, drying ovens, induction furnaces, vacuum chambers, dip tanks, ovens, wax injectors, treeing equipment, electric furnaces, pressure casting machines, and matched dies. Inspection equipment frequently employed to check the products of these processes includes radiographic X-ray equipment, fluorescent penetrant detectors, magnaflux equipment, micrographs, ultrasonic test equipment, chemical analysis equipment, and hardness testers.

• Metalworking and metal forging generally employ such equipment as hydraulic drop hammers, decarbonizing equipment, vacuum induction equipment, ring-forging screw presses, electro slag remelt equipment, and plasma guns. Extrusion machinery includes billet extruders, in-line pressure extruders, and cutoff machines. Inspection equipment used to verify the quality of products manufactured by these processes includes magnaflux equipment, die penetrant detection equipment, standard mechanical inspection equipment, microstructure viewing equipment, tension testers, and spectrophotometers for grain flow, creep, and rupture testing.

• Metal removal processes utilize a broad range of machinery, including jig borers, broaches, gear cutters, chemical mill tank lines, drill presses, grinders, lathes, mills, planers, saws, flame cutters, laser cutters, and water jet cutters. Inspection equipment used to verify the quality of products manufactured via these processes includes standard mechanical inspection equipment, hot eddy current inspection devices, chemical lab equipment, magnaflux equipment, CMMs, etch inspection equipment, and dye penetrant tanks and lights.

Ceramics Conversion Equipment

- Ceramic extrusion utilizes hydropresses and dies, drying racks, and firing furnaces. Ceramic flame spraying uses power-fed plasma arc guns.
- Ceramic injection molding uses screw injection machines, drying racks, and firing furnaces.
- Machining of ceramic parts is done with abrasive machining equipment, hydrodynamic machining equipment, abrasive jets, ultrasonic abrasive machining equipment, chemical milling equipment, electrochemical machining equipment, electrical-discharge machining equipment, electron beam machining equipment, laser beam machining equipment, or ion beam machining equipment.
- Metalizing of ceramics involves the use of plating tank lines, vapor deposition equipment, and plasma spray equipment. Pressing of ceramics involves the use of hydraulic presses and drying racks. The parts are typically inspected using tensile and compression testing equipment.
- General ceramic firing is done in ring ovens or kilns, with ceramic sintering done in sintering ovens. Slip casting of ceramics employs slip molds, clay slurry tanks, mixers, drying racks, and firing furnaces. Inspection of parts for quality through the application of these processes involves the use of standard mechanical inspection equipment, surface finish analyzers, and chemical analysis equipment.

Electronics Manufacturing Equipment

- Electronics equipment manufacturing processes are also varied and numerous. Machinery used in this field includes cast and roll machines, hand soldering equipment (torch irons and electric irons), cascade (or wave) soldering equipment, coextrusion and coating equipment, manual and automated component insertion machines, board jigs, orbital grinders, wet sanders, chemical etching equipment, photolithography equipment, finish grinders, flat draw equipment, fluxing equipment, molten tin tanks, glass melt furnaces, and annealing ovens.
- Additional equipment includes deep draw presses, polymer dispensers and cure ovens or UV chambers for plastic-encapsulated electronics, orbital polishers, sleeving machines for shielding, solderless wrap equipment, stitch wiring equipment, and surface mount equipment. Inspection equipment used to verify the quality of parts produced via these processes includes continuity checkers, laser interferometers, refractometers, chemical and optical analytical equipment, microscope comparitors, X-ray equipment, magnetic test equipment, and optic gauges.

Plastics Conversion Equipment

In the field of general plastics processing, there are several major processes that utilize specific equipment types, including the following:

- Injection molding, which uses injection molding presses, bulk material dyers, and mold shuttles
- Extrusion and coextrusion, which use extrusion screw machines
- Transfer molding, which uses hydraulic presses
- Thermoforming, which uses vacuum tables and shuttle ovens
- Blow molding, which uses blow molding machines, slitters, cooling towers, and take-up reels
- Rotational molding, which uses gyratory ovens and charge scales
- Casting, which uses mixing machines, mold-handling fixtures, and ovens

In addition, the plastics field utilizes calendering machines, plastic-blanking presses, cutoff machines, film-casting equipment, bulk material dryers, heat sealers, mills, lathes, water jet cutters, routers, saws, drills, boring machines, ultrasonic staking and welding equipment, solvent applicators, adhesive applicators, and hot stamping presses.

Composites Manufacturing Equipment

A field closely related to plastics conversion equipment is that of composites. There are two basic stages of composite manufacturing, with different types of equipment used in each.

1. The first stage is manufacture of the broadgoods composite materials. In this stage resin-impregnated fiber is used to produce composite parts in the same way coil steel is used to produce metal parts. Typically, the product of this stage is rolls of material called "prepreg," because it is a kind of fiber cloth (either unidirectional or woven) impregnated with resin. Resin formulation usually involves pumps, scales, mixers, chemical analysis equipment, splitting towers, pot reactors, high-pressure reactors, process monitors, and fraction towers. Another popular configuration of the material at this point is a much narrower "tape" used to form more radical contours in surfaces. The equipment commonly utilized in this stage includes hot melt "preggers" (or impregnating machines), solvent bath preggers, drying towers, take-up machines, tow collimators, resin-formulating equipment, and slitters.

2. The second stage involves the actual production of parts from the impregnated broadgoods, tape, or fibers. There are a number of manufacturing alternatives available at this point, including manual lay-up, automated tape lamination, filament winding, pultrusion, matched-die molding, resin transfer molding (RTM), and spray-up methods. Among the major types of machinery involved in these processes are hydraulic presses, plaster splashes, metal and composite lay-up tools, elastomer insert tools, bagging systems, autoclaves, ovens, injection molders, pultrusion machines and dies, resin injection machines, and machining equipment (e.g., routers, water jet cutters, diamond wire cutters, laser cutters, conventional mills, lathes, circular saws, and abrasive shapers).

Composite parts often require various types of nondestructive testing, including radiography, ultrasonics, acoustical ultrasonics, acoustic emission, thermography, optical holography, and eddy current. These techniques require X-ray equipment, automated C-scan or L-scan equipment, heat chambers or lamps, laser photography equipment, electrical current induction equipment, and magnetic detectors.

Metal-Treating Technology and Equipment

Metal-treating technology will remain an important part of American manufacturing technology throughout the 1990s. Among the specific processes involved are alodine, anodizing, chemical etching, dips (standard and electrostatic), spraying (standard and electrostatic), oxidizing, and plating. The machines involved in these processes include tank lines, dip tanks, drying ovens and conveyors, spray guns (standard and electrostatic), air scrubbers, air compressors, powder paint tanks, bake ovens, electrostatic generators, paint booths, sprayers, and paint pots. Quality verification equipment includes paint peel testers, aging test equipment, color meters, and coating thickness testers.

Optics Manufacturing Processes and Equipment

Optical materials most commonly are comprised of glass. Glass components can be produced by a number of different processes, including pressing, blowing, drawing, or rolling a hot, viscous homogenized melt of silica (silicon oxide), calcium, and sodium. Glass types with different properties have been developed for different applications over the last several decades, including the following:

- *Silica glass*—for applications requiring high temperature, high strength, or high chemical resistance
- *Borosilicate glass*—for applications where low thermal expansion is required
- *Lead glass*—for radiation-shielding applications or high levels of workability
- *Soda-lime glass*—for economy and workability applications such as plate glass and light bulbs

The manufacture of glass commonly utilizes specific technical processes that involve specialized equipment. The specific processes include float processing, flat drawing, cast and rolling, pot processing, blank acquisition, rough grinding, curve generation, finish grinding, polishing, and edging and chamfering. Machinery typically involved in these processes includes molten tin tanks, glass melt furnaces, annealing ovens, melt furnaces, high-purity crucibles, orbital grinders, orbital polishers, and wet sanders.

Optic fibers require some unique processes as well, including spinning, fiber annealing, etching, coating, and shielding. Equipment used for these processes includes blank furnaces, spinning machines, take-up/spooling machines, tow-bundling machines, surface-coating tanks, tunnel ovens, chemical etch cells, coextruders, and sleeving machines.

Quality verification equipment used in glass-processing environments includes chemical analysis labs, mechanical properties labs, refractometers, optical labs, laser interferometers, comparitors, tensile testers, and hardness testers.

Fastening and Joining Equipment

Current methods of fastening materials and parts together have evolved a great deal with the application of new technologies in individual fields. However, the fields themselves and the underlying concepts that define them have remained largely unchanged since the time of their initial discovery and development. The fields include heat-based fastening, adhesive bonding, pressure-based fastening, and mechanical fastening.

- Adhesive bonding utilizes application equipment, adhesive flow metering equipment, bond fixtures, clamps, ovens, and autoclaves. Quality verification equipment associated with adhesive bonding includes tensile strength test equipment and ultrasonic scanning equipment.
- Heat-based fastening includes arc welding, brazing, deep brazing, electron beam welding, furnace brazing, gas welding, heat sealing, induction brazing, jig welding, laser welding, metal inert gas (MIG) welding, pencil soldering, resistance brazing, resistance welding, traditional iron soldering, spin welding, thermal welding, thermite welding, torch brazing, torch soldering, and ultrasonic welding. The equipment used in these processes includes oxy/acetylene torches, heat sealers, induction furnaces, tungsten inert gas arc welders, lasers, MIG arc welders, high-amp power supplies, spot welders, seam welders, soldering irons, spin welding lathes, thermite crucibles, and ultrasonic generators with welding horns. Quality verification equipment utilized for heat-based fastening products includes die penetrant equipment, portable X-ray equipment, particle fluorescent penetration equipment, magnaflux, and ultrasonic evaluation machinery.
- Pressure-based fastening processes include riveting, crimping, press fitting, staking, and swaging. Equipment utilized in these processes includes rivet guns, crimping equipment, arbor presses, bucking plates, staking equipment, and swaging equipment.
- Mechanical fastening involves screwing, threaded metal insert installation, and drive pin installation. The equipment used typically includes assorted hand tools (both powered and manual), torquing tools, lifting and positioning equipment, clamping devices, and built-for-application jigs and holding fixtures.

Toward a New Manufacturing Paradigm

Current development activities in manufacturing equipment revolve primarily around improving and combining the technologies discussed in this chapter. This will continue, of course, and yield such important improvements as laser welding through flexible optic fibers, as described later in Exhibit 8-1. However, the most interesting and fundamental improvement in this realm will involve a change in paradigm. The new paradigm will not only allow us to do what we're doing today better, but will also help us to do it simultaneously, and in many cases, do it in simulation and virtual reality.

Although we see striking evidence of the advent of automation in many of these areas, it is really only beginning. As we review the relationships between design equipment, tooling, production equipment, and inspection/test equipment, a murky outline of the future starts to emerge. Especially as we move away from traditional metalworking and into the newer areas of plastics, ceramics, and optical materials, the lines of distinction between this equipment begin to blur. Automation has brought us to the brink of merging this equipment through "electronic design fixtures," testing through computer simulation, and a host of intelligence-based, machine-performed processes.

The Machines Used in Distribution

The machines used in distribution, like those described in the discussion of production equipment, fall into two broad categories: distribution logistics and inventory management equipment, and equipment for handling physical inventory.

The Automated Warehouse

Distribution management equipment, like so many other segments of the manufacturing system, has been immensely improved with the introduction of computer technology. Logistics management software packages now suggest the most efficient inventory levels to be held at the point of sale, intermediate distribution centers, and the manufacturing finished goods storage area. They also suggest the most efficient shipping lot sizes and shipping intervals. They even evaluate the fuel efficiencies, vehicle maintenance requirements, turnaround times, and distances of distribution routes to identify optimum deployment of the distribution fleet.

The machines used in distribution.

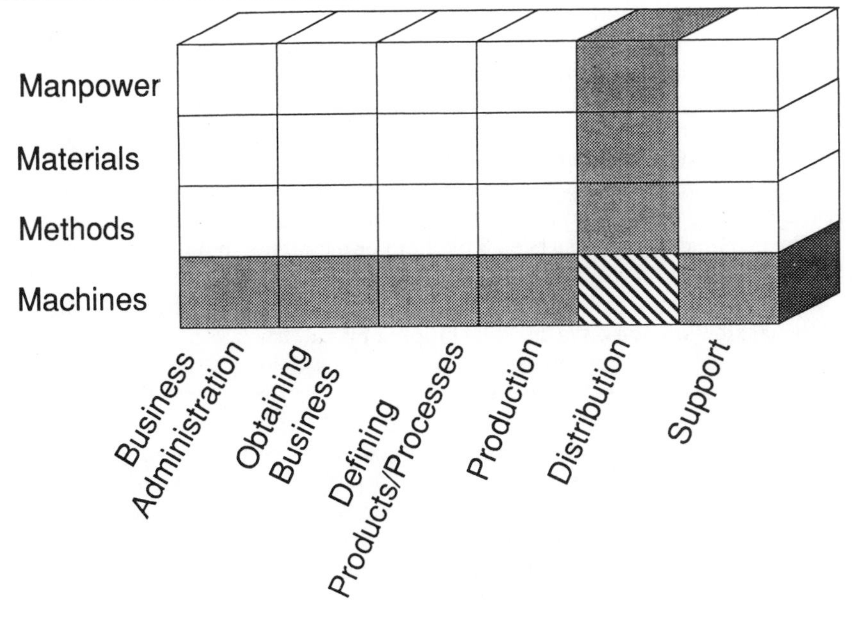

All of this is made possible by use of individual computers tied together via proprietary and public networks, through the application of electronic data interchange (EDI). This means, of course, the utilization of computers (mainframes, minis, and micros), modems for telephone hook-up, and printers for hard copies.

In addition, there is now available a host of direct-input data entry equipment for tracking parts and finished goods as they are processed through distribution and delivery. Among mainstream American manufacturers' distribution systems, it is common to find these devices at work. They include bar code readers and scanners (both fixed and portable), optical character readers (OCRs), bar code and OCR printers, data scanners, mag stripe readers, and mobile (portable/cellular) telephones. Standard office equipment (such as fax machines and copiers) are also a normal part of the distribution and warehousing environment. Also commonly used are aperture card readers and blueprint machines for verifying part identity and, less commonly, CAD vehicles (CRTs that display drawings or 3-D models electronically), for the same purpose.

Physical inventory handling has grown over the last three decades into a fairly mature automated material handling (AMH) function in most major centers of distribution and warehousing. AMH includes conventional material and parts conveyor systems, automated guided vehicles (AGVs), automated storage and retrieval systems (AS/RSs), and automatic monotractor systems (AMSs). Most of these technologies really began to flourish because of improvements in their practicality and affordability with the advent of the programmable logic controller (PLC) in the 1970s.

Lean, Green Machines

Packaging equipment itself has made great strides over the last two decades. Banding equipment has become automated and utilizes both metal and plastic banding materials. In addition, shrink-wrapping has become available and is now widely applied to fully loaded pallets of boxes. Containerization has been developed to be not only more efficient in terms of the cubic space it requires but also more versatile, often serving as a display at the point of sale. In addition, packaging has become more durable and attractive while becoming lighter through the application of modern materials technology. In the middle and late 1990s, a major challenge will be faced by the packaging industry in dealing with disposal, recyclability, and/or biodegradability as a result of our current emphasis on ecology. The response to date has been a plethora of packaging that reads "recyclable" or "produced from recycled paper."

The vehicles of distribution themselves have changed primarily in efficiency over the last few decades. We are still shipping by rail, truck, air, and ship. But the efficiency in terms of delivery cycle time, pounds delivered per gallon of fuel consumed, and cubic feet of material delivered per dollar are all dramatically improved from the 1960s.

Of course, all of this new equipment has come at a cost. Distribution activities and equipment now represent more than 30 percent of manufactured product costs in many industries, even after accounting for documented efficiency gains. The costs have seriously affected independent distributing contractors as well. Among the concerns most commonly cited are the impacts of deregulation, tax revisions pertaining to updating equipment, and escalating insurance costs. Additional factors are the continual increase in traffic congestion and the deterioration of American highways. Distribution equipment efficiencies will continue to contribute toward offsetting these costs, but it is doubtful that enough efficiencies can be gained in equipment to offset the sum of the competing costs.

The Machines Used in Customer and Product Support

The machines employed in customer support activities by American manufacturers may be categorized into communication equipment, customer- and product-tracking equipment, training equipment, and product service equipment.

Communication equipment includes not only the equipment used for normal communications with customers (telephones, fax machines, modems, printers) but also the equipment used to produce technical publications, assembly and use instructions, owner's manuals, and product service bulletins. A real boon in this area over the last decade or so has been desktop publishing, which supports the rapid development of professional-looking technical publications by almost any computer-literate customer support person. This is yet another area where vastly improved computer technology has been beneficial. A personal computer, appropriate software, and a laser printer are all that is required. In-house, communication improvements have also made customer service more accessible. Cellular telephones and pagers mean that the customer service and product support fleet is never out of touch, no matter where they are.

Customer- and product-tracking equipment has taken a quantum leap in terms of equipment sophistication and capability over the last twenty years. Many

The machines used in customer and product support.

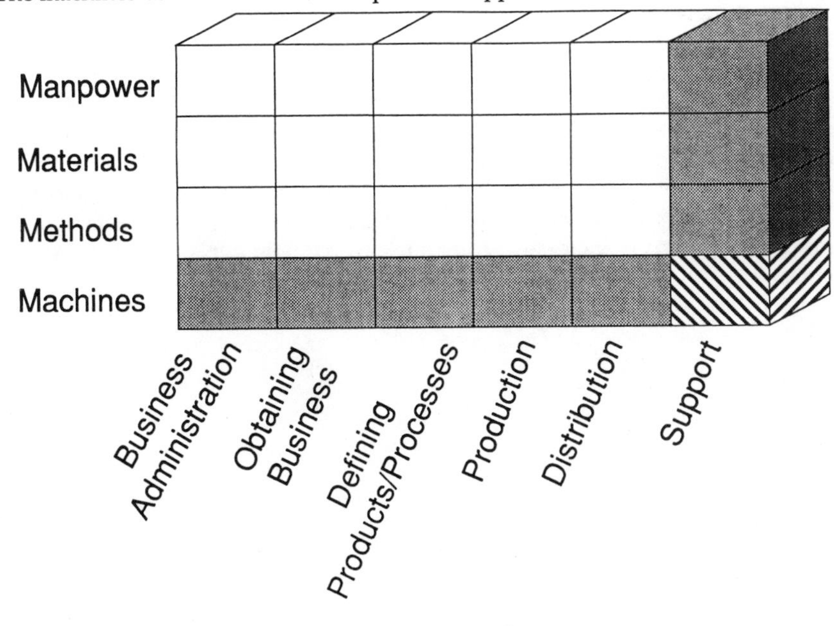

customer support organizations benefit from data accumulation devices built into the product itself, such as on-board computers in aircraft and ships. Point-of-sale product-tracking data accumulated through bar code readers can be linked to biographical information from the credit cards of consumers to identify important sales trend data and market profile data, which assist in the appropriate geographical deployment of customer support resources. Similar use of these devices results in more accurate and effective historical and predictive data regarding repairs and merchandise returns as well.

Training equipment has made real strides since the 1970s, particularly as a result of the progress made in the areas of video and computer simulation. Video-based training has become extremely popular for many kinds of manufactured products and services. It seems destined to continue to grow in popularity, because the price of developing this kind of training aid has plummeted in recent years. Also, most people simply relate better to visual images and conversation than they do to printed material. Computer simulation has brought training for complex equipment such as aircraft piloting to almost unbelievable levels of realism. As we move into the age of virtual reality, even more staggering progress will be made in this area.

Actual product service equipment has improved in two fundamental areas: the tools used for diagnosing problems with the product, and the tools used for actual product service. Diagnostic equipment is more sensitive and capable in many cases because the equipment developed over the last several years increasingly contains intelligence of its own. In addition, it is increasingly common to have technical experts actually monitor the diagnostic equipment from a remote location at the factory or regional service center and to direct the appropriate repairs from there. The application of expert systems in this setting is also a predominant feature of more sophisticated diagnostic equipment, largely because the products themselves are more complex, often containing electronic "brains" themselves, and the more complex the products become, the fewer are the people with adequate "natural" intelligence to diagnose and repair them.

The actual repair tools are often smaller, lighter, and stronger than their predecessors. In most cases, this corresponds with changes made in the products themselves. Aside from the obvious advantages of weight and size in terms of portability, the tools are easier to use for a workforce with a growing female population. Products have generally become more modular, which means that the volume of tools and variation in tool types required for most service operations is reduced, since repair increasingly amounts to modular replacement of components.

Chapter 8

Back to the Model

In Chapter 1, we briefly examined the basic components and structure of the manufacturing model. Using the manpower, material, methods, and machine aspects of each major manufacturing process, we then "broke open" each segment of the model to examine its contents. Specific elements of the current environment in mainstream American manufacturing have been described so as to build a foundation for understanding anticipated trends of the elements themselves, as well as the interaction of these elements.

To understand how the interaction of model elements and environmental factors may shape the future of American manufacturing, we need to recognize the mechanics of the interactions. As with any model, some simplification is required to convert the infinite possibilities of reality into discernible patterns and predict quantifiable results. This means, of course, that the model is an imperfect reflection of reality. In addition, there is always error involved in forecasting. When the differences between the world as it exists in a model and the real world are multiplied by normal forecast error, it is often difficult to be optimistic about the results. The degree of forecast accuracy is, of course, directly related to the value of the forecast.

How accurate, then, does our forecast need to be? The answer to that question lies in how we plan to use the forecast. A model such as ours can reasonably be expected to provide value in the following ways:

- Provide a framework for the systematic examination of as many elements as possible that may affect the future of manufacturing.
- Provide a means by which to systematically examine the elements of a specific type of manufacturing.
- Provide insights into potential developments within manufacturing from existing trends and convergence of multiple model elements. (Obviously, the more accurately we analyze the combinations of the model elements, the more accurate these insights will be.)

The model allows us to organize and systematically evaluate a great deal of fairly complex and interrelated data and turn it into useful information. It is a tool to improve management decision making by clarifying and projecting trends and relationships. That kind of insight is invaluable, almost impossible to teach outside the context of a model, and worth its weight in long-term return on investment (ROI).

Applications of the Model

How can such insights be used to improve our competitive position, and hence our individual and collective profitability? Consider these applications:

- *Strategic planning*—identifying market opportunities and anticipating changes in the markets currently served. Identifying optimum plant locations, forthcoming workforce changes, and probable regulatory developments.
- *Product planning*—identifying the attributes of future products that will satisfy needs not yet defined by the market itself. This ability puts manufacturers in the same enviable position as Sony found itself in with the Walkman several years ago.
- *Process planning*—identifying which processes must be improved or developed—and how—to meet anticipated changes in product and market characteristics. Especially in tight economic circumstances, this ability is vital when we need to determine how to deploy scarce resources.

In essence, the ability to gain insights like these translates into beating competitors to market, manufacturing at less cost than competitors, and even identifying market requirements before the market itself knows about them.

There are, of course, risks as well. Poorly constructed and poorly analyzed models generate poor forecasts, and hence a faulty foundation for management decision making. Some of the best-understood challenges associated with all types of forecasting include these:

- The ability to accurately forecast is inversely related to the time span of the forecast. In other words, the further into the future we try to predict, the less accurate we will always be.
- The more technology-intense an industry is, the more difficult forecasting will be. In most of these cases, the technologies involved are highly proprietary, and so the accuracy of current position and overall direction is more difficult to achieve.
- The more open entry is into an industry, the more difficult it is to achieve forecast accuracy for that industry. New competitors and the demise of existing competitors can dramatically alter the landscape from a market perspective.
- Demand elasticity is inversely related to forecast accuracy. That is, the more elastic the demand for the products involved, the harder it will be to accurately forecast for that industry. It is almost always easier to predict demand and developments in staples-related industries than it is to predict them when the products involved are more discretionary in use.
- It is generally easier to accurately forecast manufacturing and marketing developments related to consumer products than those related to industrial products, largely owing to differences in availability of information between the two groups.

Inaccurate forecasts can prove disastrous when a company's future is staked on them, and similar failures on a much smaller scale have ended countless promising careers. The fact is, somewhere between 35 and 80 percent (depending on whom you believe) of new products never even make a profit, and more than half of the spending done by manufacturers on new products is for products that never make it to market. Many of these failures can be attributed to poorly projecting developments in markets served, technologies used, and products fielded. Stories abound, but for every story like the Ford Edsel and the DeLorean, there is another one that exemplifies resounding success. Gillette's razor blade, Henry Ford's horseless carriage, transistors, personal computers, microwave ovens, and photocopiers are all such stories.

But accurately projecting developments in technology, markets, or specific product areas is only half the battle. The other half is convincing a financial and/or managerial support structure of the potential value of such developments. Here we have been somewhat less successful. Consider videocassette recorders (VCRs), the Sony Walkman, and the story of Nike jogging shoes. All of these products were developed only because someone believed strongly in the value of the product in the face of tremendous resistance from their peers, managers, and subordinates. VCR technology, although originating in the United States, was developed in Japan for that very reason. What a loss for American workers!

Here again, a model can provide a solid foundation for not only projecting these developments but also gaining insights into their potential value. Next we consider how this can happen.

Recall that our manufacturing model is composed of a series of component "cells" determined by major business function and whether we are studying manpower, methods, materials, or machine development.

With anticipated developments defined in each cell, it becomes possible to identify and evaluate potential consequences of these developments by examining relationships within and between the cells. There are two basic kinds of potential impacts: simultaneous convergence and sequential convergence.

Simultaneous convergence is a term that describes a rapid development caused by the impact of one or more model elements on another element. This type of change occurs quickly, almost like spontaneous combustion. This situation can occur between cells, between a cell and an environmental force, or even within a cell. Two examples of simultaneous convergence, one intercell and one intracell, can be seen in Exhibit 8-1.

The first example, fiber optic laser welding, demonstrates the rapid combination of an emerging method (laser welding) from the Production–Method cell of the matrix and a new material (fiber optics) from the Production–Materials cell that produced an entirely new manufacturing technology.

The second example, plastic fiber optics, illustrates the same principle within a single cell of the model: Production–Materials. An advanced polymer and fiber optics are integrated to produce a hybrid fiber made of plastic. The hybrid fiber can withstand heat while reducing weight and improving communications efficiency through vastly increased signal volume and quality.

Sequential convergence is fundamentally the same as simultaneous convergence with the exception of timing. With sequential activities, typically one or more of the components matures more slowly or develops significantly later than another. Exhibit 8-2 depicts two examples of this type of development.

In the first example, two relatively new developments (composites from the

(*Text continues on page 104.*)

Exhibit 8-1. Two examples of simultaneous convergence, 1990s.

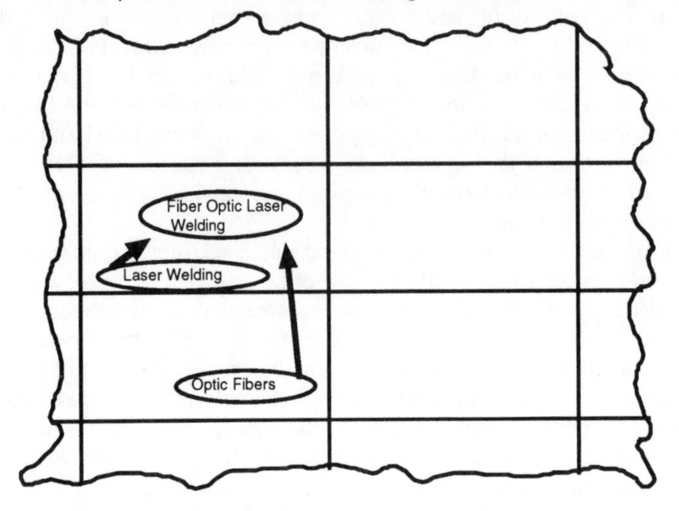

"Now researchers at GE's research and development center in Schenectady, N.Y., have found a way to "inject" a 3/4 inch laser beam into one end of a thin optic fiber with 90 percent efficiency. The new welding process produces a symmetrical weld...."

- from MECHANICAL ENGINEERING MAGAZINE, May 1989

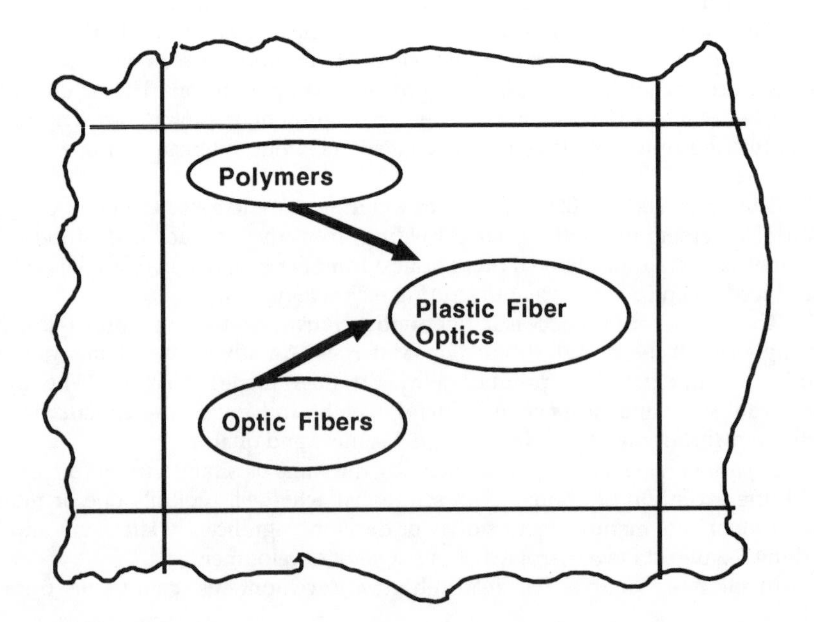

Exhibit 8-2. Two examples of sequential convergence.

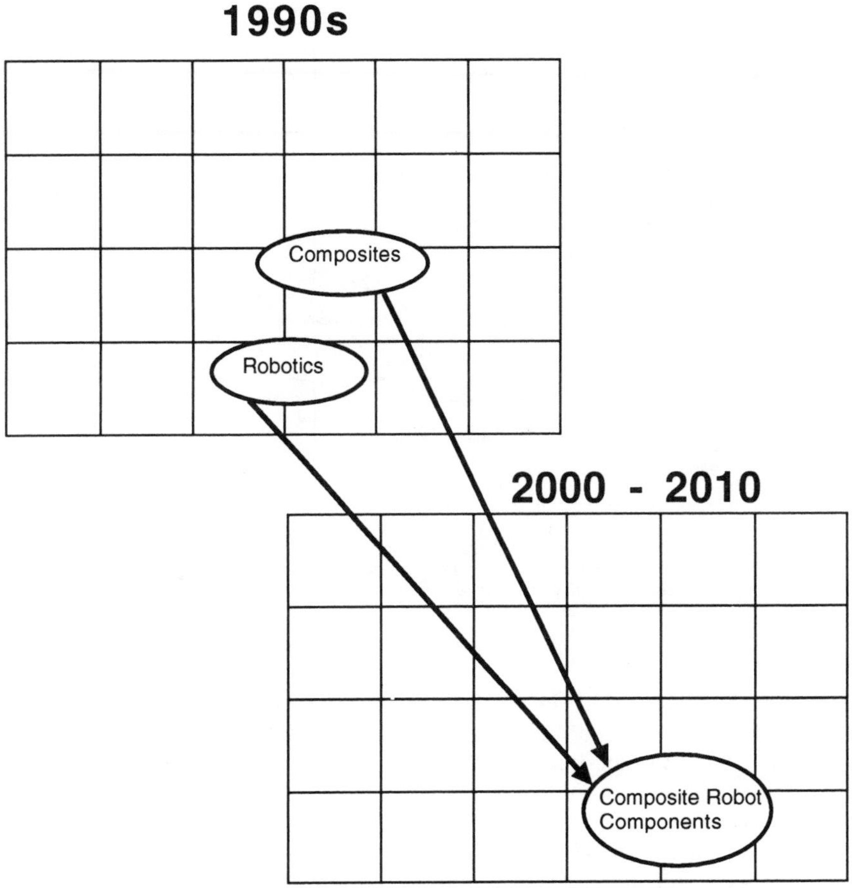

"Structural stiffness and rigidity are important factors in robots and in high-precision devices. Here the development of high strength lightweight composites, such as monofilament carbon reinforced plastics, is an emerging area of great potential. A number of robot manufacturers both in the United States and in Japan are experimenting with such materials. Mitsubishi, Shinmeiwa, and Hitachi have all reported the construction of complete arms."

- from *Mechatronics* by V. Daniel Hunt, Chapman and Hall, 1988

(*continues*)

Exhibit 8-2. (continued)

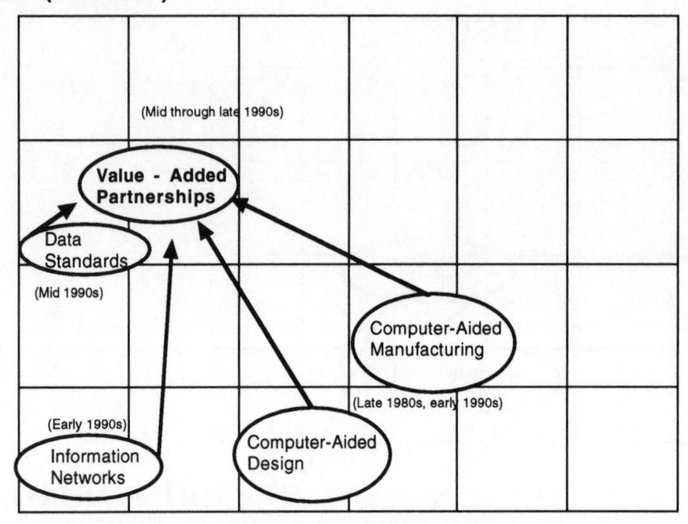

Microcomputers with user-friendly languages, inexpensive software, data standards and bar codes, computer aided design and manufacturing, and unprecedented information network integration all are converging in the 1990s to enable value-added partnerships to flourish.

Production–Materials cell and robotics from the Production–Methods cell) promise to produce a new type of component for robots over the next several years. Although still in the experimental stages, composite components for robots are an area of great potential.

The second example, value-added partnerships, required several components to be developed over a decade. Those components include minicomputers and PCs, data standards and bar codes, information networking capability, computer-aided design, and computer-aided manufacturing. These sources span the following cells: Business Administration–Machines, Business Administration–Methods, Defining Products and Processes–Machines, Production–Machines, and Defining Products and Processes–Materials.

Risks Involved in Forecasting

Applying our guidelines about the risks associated with forecasting in these cases, we could surmise the following:

• Sequentially converging phenomena will be more difficult to forecast than simultaneously converging phenomena, because sequential convergence occurs over a longer span of time. Recall from the first section of this chapter that the ability to forecast accurately is inversely related to the time span of the forecast. Although there are exceptions to every rule (including this one), it is generally correct. Therefore, we would expect that the examples in Exhibit 8-2 of composite robot components and value-added partnerships would be more difficult to predict than the examples pertaining to fiber optic laser welding and plastic fiber optics.

• Forecasting the development of value-added partnerships would be at least doubly as difficult as forecasting for the welding and fiber optics cases. This example involved not only the element of an extended time frame (as noted in the foregoing paragraph) but also a highly technology-intense development. Data standards, bar coding, information networking, computer-aided design, and computer-aided manufacturing are all critical aspects of the development of value-added partnerships. Remember that the more technologically dense an industry is, the more difficult forecasting will be. One predictive risk factor is compounded by another in this case, and the result is a very tricky forecasting problem.

• Ease of market entry in the industries involved in these examples offers some important insights about accurately forecasting developments. More specialized markets such as optic fibers and exotic polymers generally have much greater barriers to entry than computer-aided design systems, which seem to be offered by an endless number of software houses in endless revision levels. Therefore, we would expect the accuracy of forecasting developments in such areas as fiber optic welding and plastic fiber optics to be greater than that of composite robot components. Similarly, we would expect that the development of composite robot components would be more accurately forecasted than value-added partnerships. Here we see in play another of the challenges associated with forecasting discussed earlier in this chapter: The more open entry is to an industry, the more difficult it is to achieve forecast accuracy for that industry.

• Even though demand elasticity is less dramatically shown by these examples, it still affects forecast reliability. The example of fiber optic laser welding, for example, has applications in some previously difficult-to-reach locations. This is particularly suited to a number of automotive applications. Therefore, we might expect the level of resources and effort that will be expended to develop and apply this new technology to be strongly correlated with forecast R&D expenditures and capital investments for that industry.

Furthermore, the automotive industry, although a more dynamic industry than many would have expected twenty years ago (see *The Reckoning* by David Halberstam), is still closer to a "staple" business in American society than many other industries. Consequently, we would expect yet another of the forecasting challenges discussed earlier in this chapter to be in evidence: demand elasticity is inversely related to forecast accuracy. If we believed the automotive industry to be the primary market for such a development, then, we would evaluate the level of elasticity in that industry and apply the results of that evaluation to estimate the reliability of a forecast pertaining to the new technology.

• All of the examples just reviewed would be more difficult to accurately forecast than similar developments applied directly to consumer product markets and related examples. The challenge faced in these examples occurs because, as noted earlier, it is generally easier to forecast manufacturing and marketing developments related to consumer products, and we can see that these examples all fall into the "more difficult to forecast" category.

Bearing in mind these examples and the challenges to reliable forecasting, then, we now move into the world of the future.

BOOK TWO:

WHERE WE ARE HEADED

Chapter 9

Dominant Themes: From the 1990s Through the Twenty-First Century

I'll never forget my first driving lesson. Bracing myself in the backseat, I watched a fellow student jerk the car wildly from left to right, trying desperately to keep us between the center line and the edge of the pavement. Finally, as the student was about to level a stop sign, the driving instructor applied his special brake and brought us to an abrupt halt. He turned calmly to the student and said:

> Let me tell you the secret of driving. Stop focusing on the road twenty feet ahead of you. Lift your eyes, and look at the horizon of the road. Drive us toward our destination, not just down this particular section of pavement.

It was as though someone had cast a spell over the car. Immediately, the drive became smoother, with very few and much milder course corrections.

During my years as a consultant, when I visited troubled companies, I witnessed similar severe and rapid course corrections. Management changes and desperate cost-cutting gyrations sent shock waves through the companies, shaking stockholder confidence and decimating employee morale. Fortunately, in many cases, a visionary leader took charge and focused everyone on a specific, positive vision of the company's future and drove them all in that direction. The emphasis shifted from cutting to growing, and everyone was galvanized into action.

American manufacturing faces a similar opportunity. In fact, it is in the very same situation on a larger scale. We must force ourselves to lift our eyes and drive toward the horizon. Our very survival depends upon it.

Trend Spillover From the 1980s

It is important to recognize that the dominant themes of the 1990s will continue to be brought about by forces that in most cases are already in place. Therefore, we should be able to tell something about them from our previous observations.

We know this because so many of the themes of the 1980s were reactive themes. They were responses to the increasing pain of foreign competition. Most of American industry began to focus in earnest on cycle-to-market and total quality only when competition made it impossible to ignore them any longer. Just-in-time (JIT), statistical process control (SPC), cellular manufacturing, and other themes we have embraced were efforts to emulate the successes of Japanese and European manufacturers. As discussed in Chapter 2, even when the techniques were developed by Americans, the Japanese were often the first to successfully employ them.

Among the reactive themes of the 1980s that continue to spill over well into the present decade are quality, speed-to-market, automation, and employee involvement.

Quality will remain a powerful theme throughout the 1990s and the rest of the foreseeable future, because it has become evident that the marketplace will no longer be forced to live without it. Whether American manufacturers deliver it or not, our competitors will. If our products offer competitive levels of quality, they will have met that criteron required to be *considered* by the market. It's like qualifying to get into a race.

High-quality products and services by American manufacturers will be guaranteed only when we accomplish the following:

- Achieve a clear definition of the customer's needs. We must ensure that we nurture an effective process of identifying and defining customer needs even before the customer does, and then satisfying them. (Sony's Walkman is a recent example of how this has been done. Henry Ford's Model T is an American example.)
- Standardize efficient, error-free performance in our daily operations. We must identify which characteristics of our products and services are important to the customer, and which internal processes generate those characteristics. Techniques like SPC may then be applied to the critical processes to proactively identify problems and prevent them. Some processes will need to be completely reengineered to meet the goal of customer satisfaction at minimal cost before statistical process controls are applied.
- Measure performance and demand continuous improvement. Metrics that reflect true customer needs and our performance in satisfying those needs must be established and linked directly to the compensation of everyone, from the assembly line worker to the CEO.

Speed-to-market is another reactive theme of the 1980s that is likely to grow in importance over the next decade. Robert Gilbreath in his book *Forward Thinking* (McGraw-Hill, 1987) says:

> If an item takes too long to conceive, analyze, study, design, prototype, test and produce, it won't be approved. The bottom line still exists; it is simply moving up the page.

Joseph Vesey of Unisys Corporation agrees. In an article entitled "The New Competitors: They Think in Terms of 'Speed to Market,' " he states:

> The emphasis in manufacturing companies during the 1990s will be "time-to-market"—that is, the elapsed time between product definition and product availability.

Vesey goes on in his article to quote a McKinsey & Company study that indicates that if a company is six months late to market, gross profit potential is reduced by 33 percent.

Automation will continue to be another dominant theme throughout the decade ahead as well. It will be pervasive in every aspect of manufacturing, from simulation for executive decision making through design, production, distribution, and customer support. It will appear in a plethora of applications, from robots to flexible manufacturing systems (FMSs) to computer-aided design and computer-aided manufacturing (CAD/CAM) systems. Many mainstream manufacturers will continue to pursue the computer integrated manufacturing (CIM) approach, although experience suggests that the approach will require a different name in many organizations, since CIM is now a dated acronym and therefore hardly fashionable any longer.

Employee involvement will probably be a somewhat less heralded spillover theme from the 1980s, but will almost certainly be an increasingly important aspect of manufacturing well into the twenty-first century. Manufacturing jobs will continue to diminish in quantity in the foreseeable future, and that means more of workers' gray matter will be needed to supplant large support staffs and diminishing overhead budgets. This phenomenon will also be fueled by the increasing desire of American workers to have more meaningful work.

Bold New Initiatives

Beyond those dominant themes that are largely responsive to competition and other external phenomena, there will be some themes that dominate because we recognize their importance and proactively undertake corresponding initiatives. Among the initiatives already on the horizon are concurrent engineering, process integration, strategic alliances, and "green" manufacturing (environmentalism).

Concurrent engineering will be a dominant theme as we enter the twenty-first century because it supports so many other related requirements of an increasingly competitive global marketplace. By considering many of the subsequent manufacturing and distribution processes during the design activity, the cycle-to-market is shortened. The overall cost of manufacturing is reduced because manufacturing capabilities are considered in product design, and manufacturing processes can be defined and documented prior to design release. Supportability of products (and therefore customers) is improved because products and processes can be designed to facilitate service and overhaul. Even packaging and logistics can be considered in the design equation.

Concurrent engineering is a single example (and one of the most well-defined examples) of the larger and more pervasive theme called process integration. American manufacturers over the next decade will increasingly recognize that the "silos" of individual departments and functions must be broken down and integrated to mirror the processes that span manufacturing companies. For example, there will be less likelihood of finding a "purchasing" department and more likelihood of finding an organization that performs the entire process, from receipt of customer order through negotiation of the contract for procurement of parts and raw materials. As the fundamental processes within a manufacturing organization are better understood, documented and integrated, there will be less redundant and non-value-added activity. Competitiveness will improve.

Strategic alliances are another theme that will almost certainly dominate the manufacturing environment of the 1990s and probably stretch well into the twenty-first century. Strategic alliances allow manufacturers to share the risks of development associated with new products and services and to open doors to foreign markets through satisfying national labor content laws. When properly constructed, these arrangements enable participants to take advantage of the core competencies of other companies and produce a better product or service than would have been possible otherwise.

Environmentalism and its underlying philosophy will almost certainly grow in importance over the next couple of decades. Already the impacts of costly processes associated with the monitoring and disposal of hazardous materials have become a significant financial burden in many industries. Manufacturing faces enormous challenges in this area in terms of financial performance and regulatory compliance.

Manufacturing for the Global Market

What we are likely to find, then, in terms of overall themes that dominate the 1990s and the first decade of the twenty-first century, are manufacturing organizations with a far more global market and supplier base. They will be driven by a burning need to move innovations to market at breakneck speed, requiring risk-taking and enormous flexibility. The flexibility will be reflected in organizational structures, which will evolve to emulate the actual processes of the business rather than function as traditional organizational structures. The lines will blur between management and the shop floor as more and more intelligence is required to perform conversion processes and automation seeps into virtually every facet of the business. The thirst for information pertaining to customer needs will be insatiable as markets continue to diversify and become more demanding. Quality, the value of the product as it relates to price, and the tailoring of the product to the end consumer will often mean the difference between winners and losers in the years ahead.

Chapter 10

Environmental Factors: Building Alliances in a Fragmented World

Environmental factors (see Exhibit 10-1) as we move toward the next century will continue to be the most powerful influences of all on manufacturing. The political, economic, enabling, social, and competitive environments have shifted significantly since we entered the 1980s, and the rate at which developments are occurring is staggering. The unpredictable nature of environmental factors combined with the speed of change creates a tempest of market, supplier, and competitor influences that we must face daily.

Political Influences

As we pass through this last decade of the twentieth century, political influences promise to be a mixed blessing. Import controls, for example, which are still advocated by many of our leading political figures, protected nineteen separate industries in 1986. According to Robert Hamrin in *America's New Economy*, that particular defensive tactic affected import prices dramatically (and domestic prices somewhat less directly), resulting in a cost to American consumers of $66 billion dollars. However, the failure of protectionism notwithstanding, it is a relatively easy to implement, quick-fix solution certain to grace the platforms of ambitious politicians. Over the next twenty years, with growing foreign pressure on American industry's profitability, people will likely continue to press for it. If it dies, it will die hard.

Government as Partner, Not Adversary

The relationship between government and manufacturing companies will change significantly over the next ten years. It will have to, or most of our existing manufacturing base, already diminished, will simply disappear. Government and industry will have to become far less adversarial. With both entities suspicious of each other's motives, value, and credibility, the primary objectives of each often become "catching *them* in the act" of some inappropriate or illegal activity, keeping *them* at bay, so that *we* can get on with *our* job, and revealing as little as possible about what *we* know to *them*. Over the next ten years or so, the roles of government and industry will be forced to evolve from their current "watchdog" versus "opportunist" relationships. Government will need to develop into a coordinating,

Exhibit 10-1. Environmental influences likely to have significant impacts on American manufacturing.

Political:
- Government / Industry Alliances
- Deregulation
- Environmental Protection Regulations

Competitor:
- Shift From "Make" to "Buy"
- Consolidation Based on "Core Competencies"
- Nondomestic Product Development Content

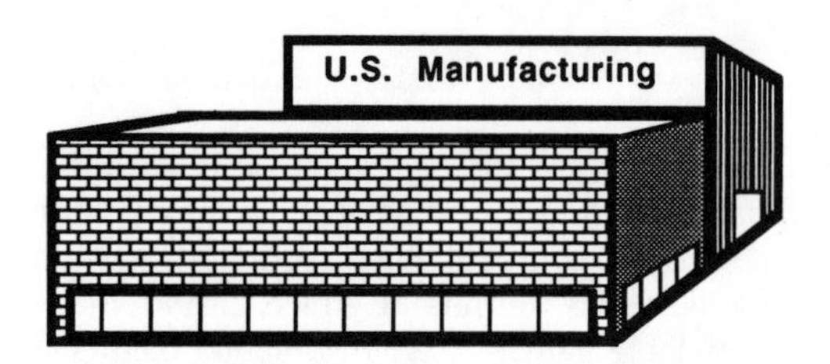

Enablers:
- Chaos Theory
- Virtual Reality
- Alternative Energy Production
- Digital Signal Processing

Social:
- Aging Workforce
- Changing Worker Values

Economic:
- Tri-polar Economy
- Fragmented Market
- Inadequately Educated Workforce
- National Deficit / Debt
- North American Free Trade Agreement

facilitating, and strategic planning entity. It must develop, communicate, and provide incentive for participation in strategic technology development and deployment. We can see important steps being taken in a very positive direction already, with the advent of "Process Based Management" (PBM) efforts. These initiatives represent an impressive and powerful "teaming" of government and contractor personnel within the aerospace and defense industry to make manufacturing processes more efficient and cost-effective.

Government must deregulate to the maximum extent possible in areas that would encourage strategic business alliances similar to the *keiretsu* business groups of Japan and similar cooperative business groups now developing in Europe. Members of these groups are often suppliers, manufacturers, and distributors who are all interconnected by the nature of their businesses. They commit to one another's success by owning large blocks of stock in the others' companies.

This approach offers protection from hostile takeovers, advantages related to centralized planning and operation coordination, and freedom from the pressure of outside shareholders to attain short-term profits. Current antitrust laws largely prohibit such cooperative business structures in the United States, as they have since the 1930s.

Government will almost certainly also recognize that it must develop a national policy related to what strategic technologies should receive preference in terms of research funding, and shift its emphasis in this regard away from military applications toward commercial ones. Even the small nation of Singapore has recently adopted a national policy of this type, calling out key technologies it will directly support. (Among those key technologies, incidentally, is materials science as it relates to manufacturing.) Consider Japan's Ministry of International Trade and Industry (MITI), which adopts strong policies that encourage product standardization, product specialization by specific company, and R&D funding and support. If America is to compete as a nation, it must coordinate its efforts through national policy to "level the playing field" with other countries who already do.

Government is likely to continue with deregulation in other areas, but the effort will be fragmented. As constituencies become increasingly environmentally aware and health conscious, regulations will be called for in increasing number, especially in those fields related to "hazardous" materials. Unfortunately, while the spirit of these activities is entirely appropriate (and in many cases long overdue), the decisions are fraught with emotion and often made on somewhat subjective grounds. Rosalyn Yalow, a Nobel-winning physicist, writes:

> Executive, legislative, and judicial branches of our federal, state and local governments need to search for more effective science and technology advice, rather than responding to the loud voices of political activists ("Over-Regulation: Required or Ridiculous?" In *Science and Technology Advice to the President, Congress, and Judiciary,* edited by William Golden. New York: Pergamon Press, 1988).

This kind of dialog is indicative of a movement toward rational deregulation that is critical to remove unnecessary constraints in many manufacturing- and distribution-related areas. The objective is not "no regulation," of course, but rather "responsible, reasonable regulation" that successfully balances the protection of environmental and social needs with the freedoms required for our businesses to compete. Nobel-winning physicists aren't the only ones who can see this problem clearly—ask any manufacturer facing the tough new environmental regulations now being levied in the United States. However, it *will* require some real moral and intellectual strength to develop and deploy this kind of regulation fairly over the years ahead.

Economic Influences

The world's economy has been evolving over the last fifty years from a single economic "pole" (the United States) toward a bipolar system (the United States and Japan), and it is about to change again. This time, the economic complexion of the world will be characterized by three "poles": a tripolar economy led by the United States, Japan, and Europe. In a remarkable book entitled *Head to Head—The*

Coming Economic Battle Among Japan, Europe, and America (Morrow, 1992), Lester Thurow of MIT states:

> Building upon the economic muscle of Germany, Western Europe is patiently engineering an economic giant. If this bio-engineering can continue with the eventual addition of Middle and Eastern Europe, the House of Europe could eventually create an economy more than twice as large as Japan and the United States combined.
>
> . . . the integration of the European Common Market, on January 1, 1993, will mark the beginning of a new economic contest in a new century at the start of the third millennium. At that moment, for the first time in more than a century, the United States will become the second largest economy in the world. This reality will become the symbol for the start of the competition that determines who will own the twenty-first century.

Our Uphill Struggle

Economically, the United States was unstoppable in the years following World War II. The United States has a market more than nine times bigger than its nearest competitor, owing primarily to superior technology emanating from intellectual giants like Einstein and Fermi, a workforce that was more highly educated and skilled than any other, the highest GNP in the world (which provided ample investment funding advantages over foreign competitors), and highly skilled professional managers.

As we pass through the 1990s and enter the next century, the United States (and U.S. manufacturing specifically) faces a whole new set of economic realities. For example:

- The United States is no longer the largest market in the world.
- Our desire for more customer-tailored and unique products and services has weakened the economies of scale once offered by mass production.
- Although still superior in many areas of *product* technology, we are quickly losing ground because of our inferior *process* technology
- We are no longer leading the world in R&D spending, and are especially poor in nondefense R&D.
- The American workforce is not adequately educated to utilize modern technologies. No amount of managerial training can overcome an inadequately skilled workforce.
- Eleven countries exceeded the United States in wages by 1990. Manufacturing wages specifically were higher in fourteen other countries. The gap between U.S. manufacturing wages and manufacturing wages in West Germany in that year was more than $9 an hour.

There is no apparent reason that these trends will do anything other than accelerate over the next decade, as the European Economic Community (EEC) solidifies and Japan continues to erode our own domestic market share in industry after industry.

The federal budget deficit will undoubtedly continue to exacerbate this situation. During the 1980s, the percentage of our federal spending allocated to the interest payment on this debt rose from 9 percent to 15 percent. We continue to spend more than we bring in, which increases this debt level. As a result, the U.S. tax system will almost certainly be overhauled. It is, of course, impossible to be certain how that will come out. However, it seems reasonable to believe, from our current spending patterns, that we will be forced to raise tax revenues to come to grips with our mounting debt. The method devised may well include some form of "value-added tax."

Whether tied to income or consumption, an increase in the tax structure will impact American manufacturing. If levied in the form of traditional income taxes, individual savings are inadvertently discouraged, since there is less disposable income. Less savings means less capital available for plants and equipment, and so on. If levied in the form of a value-added tax, the tax on manufactured products goes up, discouraging their purchase. Either way, this amounts to bad news for manufacturers, at least for several years. There are, of course, myriad opportunities to reduce spending. However, as discussed in Chapter 3, this behavior is, uh, inconsistent with our recent and current behavior patterns.

Health care costs, already a major contributor to the employment costs of American manufacturers, will continue to grow over the next decade. Among the reasons commonly cited are these:

- Liability awards for malpractice
- Malpractice insurance
- Expensive new technology
- Overpaid doctors
- Inefficient hospitals
- Excessive paperwork
- Overuse of benefits
- Unnecessary medical procedures

Health care expenditures, which were at $1,900 per capita for 1987, are expected by the National Center for Health Statistics to grow to $6,000 by the year 2000.

Unemployment levels should flatten out and begin a slow decline over the next several years, because of a significant increase in service-related jobs coupled with the reduced number of new entries into the labor force resulting from smaller families.

Unfortunately, the reduction in the labor force will not keep up with the expected reduction in manufacturing-related jobs. In a book entitled *Future Scope—Success Strategies for the 1990s and Beyond* (Longman, 1990), Joe Cappo says:

> The 1990s will develop into an unprecedented workers' market, which will put tremendous pressure on all kinds of businesses. The only industries not affected will be those that are shrinking, primarily blue-collar manufacturing operations. This sector of the economy will produce very few jobs in the next ten years.

This is probably somewhat of an understatement. Among the categories of jobs expected to experience the biggest percentage decline through the next decade are these:

- Electrical and electronics assembler—expected to lose 133,000 jobs (54 percent) between 1986 and 2000
- Electrical semiconductor processor—expected to lose 15,000 jobs (51 percent) between 1986 and 2000
- Industrial truck/tractor operator—expected to lose 143,000 jobs (34 percent) between 1986 and 2000
- Chemical equipment controller—expected to lose 21,000 jobs (30 percent) between 1986 and 2000
- Chemical plant/systems operator—expected to lose 10,000 jobs (30 percent) between 1986 and 2000
- Textile machine operator—expected to lose 55,000 jobs (25 percent) between 1986 and 2000
- Coil winder/taper/finisher—expected to lose 6,000 jobs between 1986 and 2000

Other categories experiencing similar declines include press machine operators, extrusion machine operators, drawing machine operators, and machine maintenance mechanics.

A superficial review of this data might suggest that there will be abundant skilled workers available to manufacturers to perform this kind of work, and that much is true. However, that is not good news, because the nature of work in the manufacturing environment is changing. As it changes, these skills are required less often.

The National Research Council made the following statement in a treatise entitled "Toward A New Era In U.S. Manufacturing" (Washington, D.C.: National Academy Press) in 1986:

> . . . the factory will become a much different factor in society. Although opportunities for unskilled or semiskilled labor will diminish, the jobs that will be created are expected to be challenging and of high quality. Also, manufacturing jobs will be in demand among graduate engineers, who do not generally prize them today, and there may be too few to go around.

American Demographics, in a March 1990 article entitled "Workers in 2000" written by Diane Crispell, stated:

> Technical workers are projected to be the fastest-growing major occupational group over the next decade, and the United States isn't producing enough of them.

Social Influences

Social influences, as discussed in Chapter 3, may generally be classified in two broad categories: demographics and values.

Demographics 2000

Much has been written about the aging U.S. workforce. In 1980 the median age in the United States was 30. By 1990 it was 33. By 2000 it is expected to be 36, and by 2010, 39. The population of those under 35 is projected to decline by 3 percent between 1980 and 2000, while the over-35 group is expected to increase by 46 percent.

These projections are significant to manufacturers for several reasons. One frightening implication is that retirement benefits to be paid out will increase enormously. In 1980, there were 5.5 people in the working age bracket for every individual of retirement age. By 2000, the ratio will be about 4.7 to 1, and by 2030 it will drop to a ratio of 2.7 to 1. (Between 2012 and 2030, some interesting things should happen to the tax structure if current levels of social security support are to be maintained.) The aging population is already challenging the assumptions of some companies' pension funds, which were constructed using actuarial tables projecting significantly shorter life expectancies.

New Opportunities for Women and Minorities

Two-thirds of the new hires over the next decade will be women, and about 85 percent will be either women or minorities. Generally speaking, new employees will require more support than ever in terms of education and training in an era where the sophistication level of job skills required will escalate along with the advances in manufacturing technology. Unfortunately, these new employees will not have such training. Even those now pursuing university degrees are not getting them in the fields that would be advantageous to American manufacturers.

For example, here is the percentage of university degrees earned by women in various disciplines in 1990: Women earned 60 percent of the degrees awarded in accounting, 55 percent of those awarded in business (arguably relevant), 53 percent of those in law, 48 percent of those in medicine, 40 percent of those in computer science (of some significant value to manufacturers), 35 percent of those in agriculture, 35 percent of those in architecture, and only 25 percent of those awarded in engineering (the most relevant field for manufacturers in the decades ahead).

Manufacturers, then, can expect the costs and investments required to build a world-class workforce to increase dramatically over the next decade as they pick up more and more of this cost. In addition, this drain on resources is likely to adversely affect the learning curves associated with new manufacturing technologies and hinder the required steps in reducing cycle-to-market of new products and services.

The Leisure Society: From Layoff to Time Off

In terms of their values, American manufacturing workers are in the middle of a significant shift away from the traditional "work ethic" toward what is increasingly referred to as a "leisure society." They want to work fewer hours of the week. They want work that is intellectually stimulating as opposed to dull and repetitive. This is arguably pretty well suited to the way manufacturing will develop over the next couple of decades. Much of the unskilled and semiskilled labor will disappear, and work will require more intellectual energy.

In the nearer term, how might this desire to work fewer hours affect manufacturers? Let's explore an example. Let's say you were just hired as the president of a manufacturing company that produces lawn and garden equipment, including snow removal equipment (to eliminate the seasonality problem for the sake of this discussion). The plant is currently working only one shift, and because of cost overruns on some recent development programs, there have been a series of layoffs and other cost-cutting efforts. You know there is a market for more than you are making, and you know that getting the cycle-to-market down is an important aspect of competing in this market. What do you do?

One answer might be to adopt a ten-hour per day, four-day per week workweek and then hire three times as many people, so that the plant operates twenty hours per day. Everyone works four days, then takes four days off, then works the next four days, and so on. Of course, this means people will have to live with working what are currently referred to as "weekends" at times, but think of the advantages. For what would amount to about a 6 percent reduction in pay, they would literally get half the year off! They could use their free time to continue their education, or even take a second job if they really wanted to (the vast majority wouldn't). The number of employed people would quadruple, and the cycle-to-market would be reduced by nearly continuous production. The four hours left each day would be an ideal period for preventive maintenance and general janitorial activities and would provide the needed cushion for overtime in problematic periods.

The increase in such time-saving services as professional cleaners, pet care services, and lawn care prove that time is "the currency of the future." However, the service industry isn't the only one that can benefit from understanding this trend in the American consumer, because the American consumer is also the American worker. Creative approaches in deploying the American manufacturing workforce can result in winning combinations for everyone.

Dangerous Values

The same values that have driven the rate of American spending beyond prudent limits, and transformed us from a creditor nation to a debtor nation, are also responsible for the "short-term profitability" orientation of modern American managers. This distinction means that any foreign competitor who wishes to take a market away from American manufacturing has only to be willing to endure low levels of return on investment (ROI) long enough to drive American producers from the picture. Most Japanese competitors are more than willing to make this kind of investment, which is how they happen to own so many industries that used to belong to the United States. They simply started at the low end of the market, incurred marginal ROI for a few years, and waited for American manufacturers to abandon that market segment. When we did, the prices could be (and usually were) raised by the Japanese firms, and their profitability was restored. In addition, they concentrated on making their manufacturing processes more efficient, and put a virtual lock on that market segment. Then they moved up to the next market segment and employed what they had learned to move even more quickly.

As long as manufacturing performance is measured by short-term ROI rather than market share, this situation will persist. It is a function of increasingly demanding stockholders, a function of greed.

Competitor Influences

As we enter the twenty-first century, we will find that our competitors have most of the advantages in terms of manufacturing capability. In many cases, their workers are willing to accept lower wage levels than American workers. Their factories are more modern in many cases and better equipped with state-of-the-art technology, because we have been moving an increasing amount of our manufacturing operations offshore and because foreign competitors have been developing manufacturing process capability while we have concentrated more on product development. In addition, most of the Pacific Rim competitors have lower overhead rate structures, because they maintain much flatter organizations with less nonlabor support.

What we have that our competitors are most interested in is our market. Second, we have a remaining core technology competency, primarily in terms of new product technology, although there is real momentum now to move that offshore through strategic alliances and foreign procurement of American companies as well.

Three Trading Blocks, Three Major Shifts

As we move through the late 1990s, the United States will be relegated to a position as one of three roughly equal trading blocks within a tripolar trading base (the others being Japan and the EEC). Competitors in the other two trade centers are in the advantageous position of being part of national strategies and plans. Industries and R&D are subsidized (e.g., Airbus) and orchestrated by national and regional planning groups (such as Japan's MITI). These factors, combined with the process improvements already in place within their facilities, will result in tremendous pressure on the remaining American manufacturing base.

As a result, we will see three major things happen in the next two decades:

1. Make/buy decisions over the next ten to twenty years will increasingly shift from "make" toward "buy." This will be the first clear sign of the loss of the remaining American manufacturing base. Companies that wish to be competitive producers in the long run must invest in, develop, and maintain unique capabilities for the fabrication or assembly of their product. Generic processing capabilities will always be available from other companies, and often they will be less expensive to procure. This is a function of having concentrated on products rather than processes, as discussed in the section "Social Influences." This is not a winning strategy. Consider the differences between the strategy of off-loading uncompetitive processes versus the strategy of improving uncompetitive processes, as shown in Exhibits 10-2 and 10-3.

(Text continues on page 124.)

Exhibit 10-2. How to grow a business.

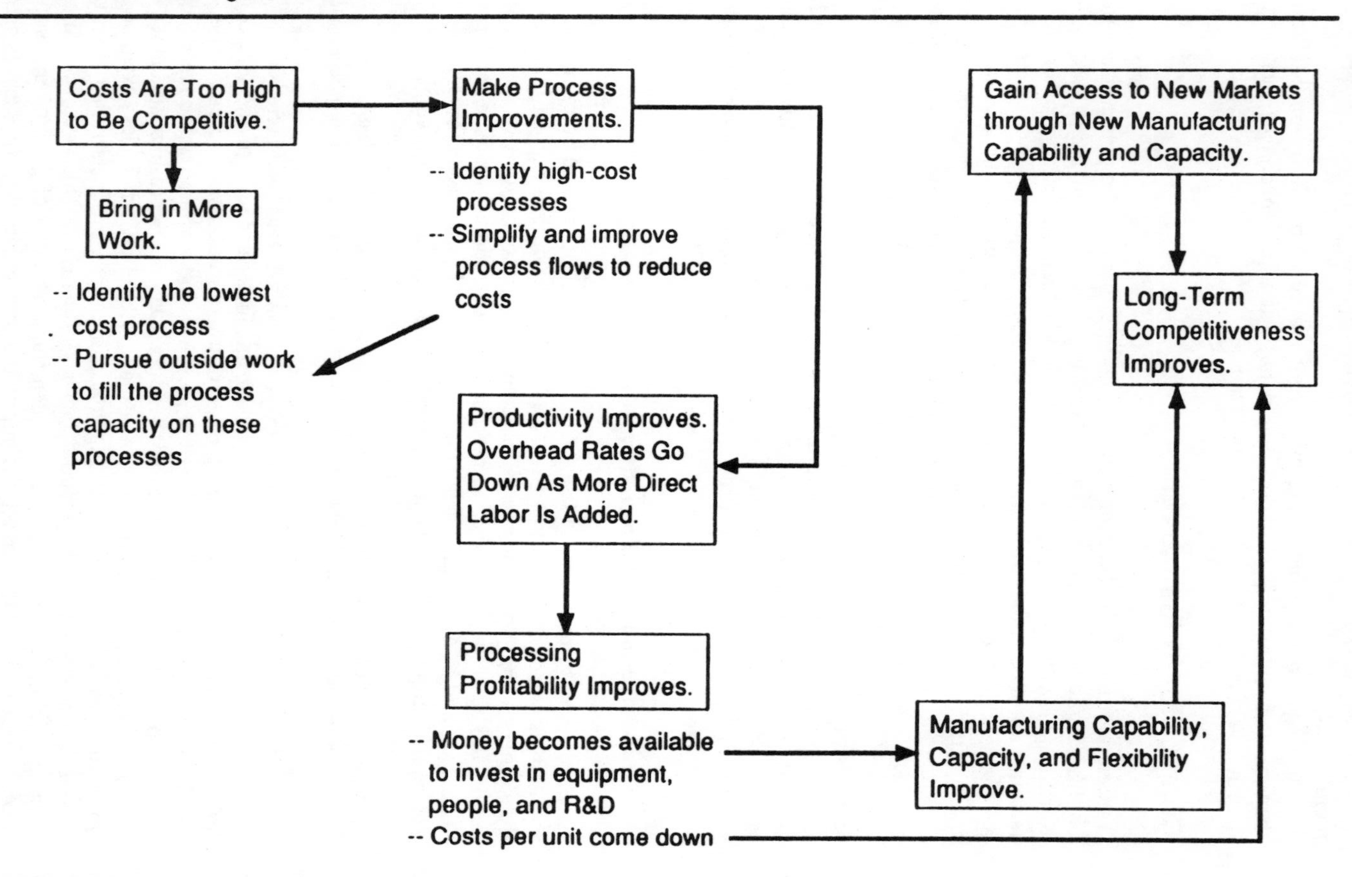

Exhibit 10-3. How to go out of business.

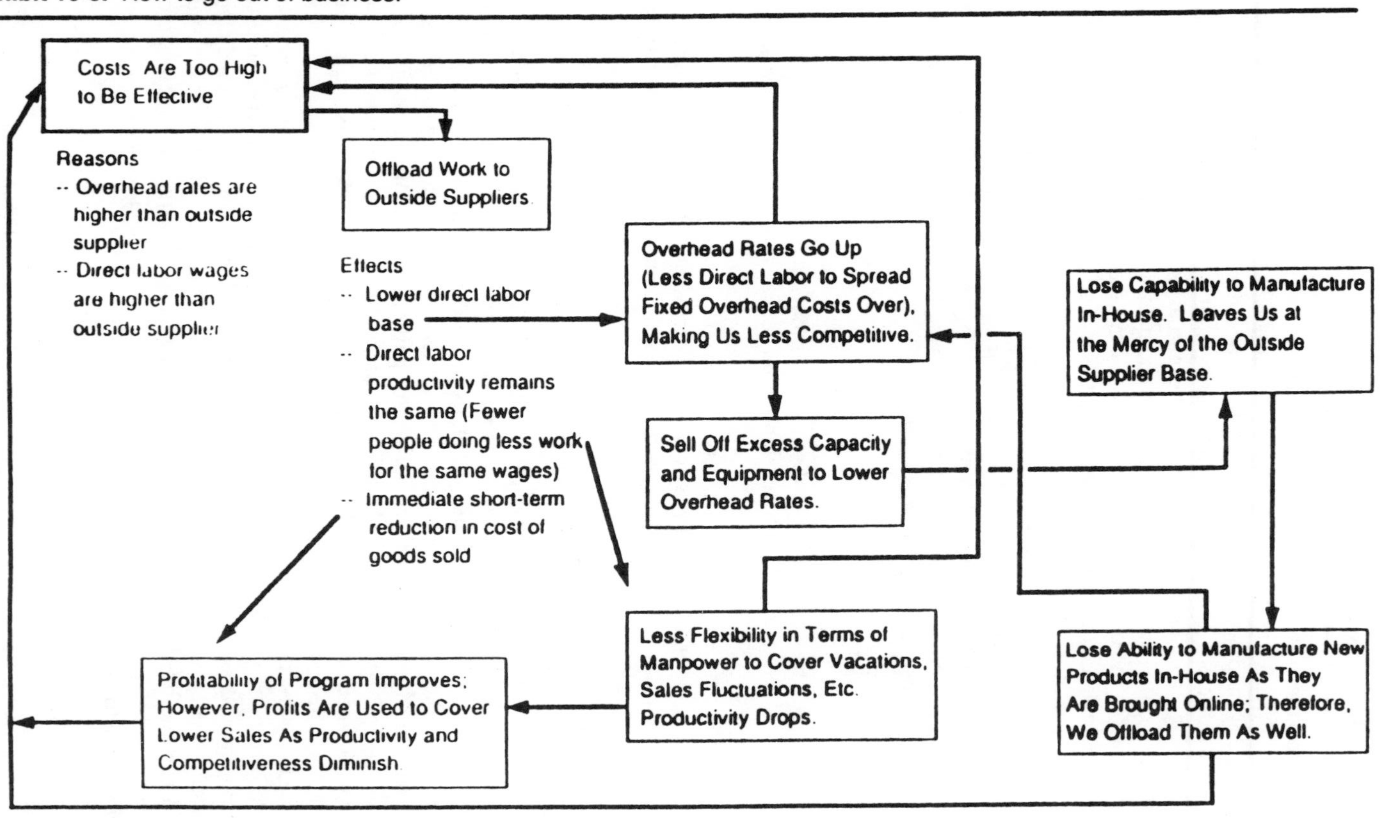

2. American manufacturers will increasingly turn to "core competencies," and find that this strategy fails. American manufacturers will make a concerted effort to reduce their own manufacturing operations to select processes where they perceive themselves to be superior. Unfortunately, truly superior and unique manufacturing processes are becoming fewer and fewer in the United States.

It is a strategy geared toward short-term profitability, which inevitably results in declining manufacturing capability and eventual extinction. It can be effective as a strategy only if there is a national strategic planning initiative that coordinates and sponsors the development of these core competencies. Thus far, there is little evidence that such an event will occur in the United States. However, we desperately need such leadership.

3. Major product developments will rarely be done entirely within the United States. "Strategic alliances" and the level of foreign ownership in U.S. manufacturing endeavors will undoubtedly continue to grow. This will be true not only because of the erosion of our own manufacturing base but also because the major sources of investment capital will be overseas.

Enabler Influences

Enabler influences are powerful discoveries, innovations, or theories that enable mankind to make significant developmental leaps, often across a broad band of disciplines.

Enabler influences, by their very nature, are virtually impossible to predict. Some of the more interesting possibilities appearing on the horizon include chaos theory, virtual reality, alternative energy production approaches, and digital signal processing (DSP).

Chaos in the Workplace

Chaos theory is a recently developed field of study that examines the complex, irregular behavior of many naturally occurring systems, such as weather patterns, the paths of billiard balls as they move about on a table top, and the movement of particles in the rings of Saturn. Scientists in this area are developing equations that define (and thereby predict) the long-term patterns of "chaotic" systems. In the same manner that quantum physicists predicted the existence of some subatomic particles based on the "probable" location of other particles, scientists study chaotic systems to measure and predict them.

This work is likely to affect biotechnology, materials engineering, and very high tech manufacturing processes. Advances in this field should yield improvements in processes that offer great potential when they become more predictable. Among them are composites autoclave processes, fluid and glass flow dynamics, and ceramic brittleness reduction. Some years in the future, this approach may also provide important insights regarding such predictability problems as sales forecasting for service parts.

The New (Virtual) Reality

Virtual reality is a field in which three-dimensional computer simulation is taken beyond its current visual limitations into visual, audio, and even tactile realms. It will allow designers, for example, to actually "see" the object of their work, and eventually manipulate it with their hands as though they were physically applying force to it.

Even now, prospective home buyers in Japan are using virtual reality to "walk through" their new homes before they are constructed, in order to make the changes they would like. Virtual reality "arcades" have been set up in shopping malls in many areas of the United States to explore the gaming potential of the new technology.

The manufacturing world we will eventually occupy will involve constructing new materials and altering the chemistry of existing materials at the molecular level. What better way to simulate the disassembly, replication, and reassembly of the molecules required than through virtual reality? When this has been achieved, applications of robotics to this technology will even take us out of the simulation world and allow us to build the new materials and their products in a similar fashion.

Harnessing Enzyme Energy

Alternative energy production will be developed in several different ways, including solar, geothermal, tidal power, wind power, enzyme bioproduction, and fusion applications. It is, of course, unclear which will be the real winners in this area, but the availability of energy will always have a significant impact on manufacturers in terms of facility locations, overhead rates, and general operating efficiencies.

Among the more interesting fields of alternative energy development is enzyme bioproduction, the process of manufacturing complex synthetic hydrocarbon chains to replace those currently produced in petroleum refining. Genetically engineered enzymes are constructed from hydrogen, oxygen, and carbon molecules and assembled into the hydrocarbon chains required.

Another promising field is fusion. Fusion involves utilizing the collision of hydrogen atoms to release enormous amounts of energy. According to some analysts, fusion is likely to become the dominant source of large-scale electrical power generation in the twenty-first century.

Digital Signal Processing

Digital signal processing (DSP) is probably as close to a specific technology as the "enabler influences" category can contain. Although it is in fact a specific technology (as opposed to a broad theoretical concept), it meets the criteria for enabler influences.

DSP is an enabler that allows continuously varying analog signals (i.e., sound or images) to be translated into the digital language of computers (1's and 0's). This is done with DSP chips, and the results affect almost every possible electronics market niche. By utilizing DSP to compress audio and video signals, for example, hundreds of cable TV channels can be brought into each home. AT&T utilizes DSP to virtually eliminate moving parts in answering machines. Manufacturers of loud equipment, from automobiles to lawn mowers, will be able to use DSP to emit noise-canceling signals from these devices, making them virtually silent.

Fortuitously, the complexity of the algorithms that make up DSP is significant, and so may require more time than most other products to "reverse-engineer." The United States is the leader in this technology at the moment, heavily supported by Israeli mathematicians. According to Gene Bylinsky in an April 20, 1992, *Fortune* article entitled "A U.S. Comeback in Electronics,"

> American DSP successes are creating new manufacturing jobs. Motorola is expanding its facilities in suburban Chicago to meet a growing demand for cellular phones. Companies that supply DSP know-how and hardware are booming.

There are bright spots, then, in the "enabler" area at least.

Summary

As we move through the 1990s and into the next decade, we face another set of powerful environmental influences. American manufacturers will have to deal with the increasing strength of global competitors, an increasingly technology-intense manufacturing process environment, a workforce motivated by "leisure society" values, a growing mismatch between the education and skills of the workforce and the needs of the manufacturers, and greater scarcity of domestic capital sources for investments in plants, equipment, and R&D. At the same time, our markets are becoming global and more demanding of products and services tailored to their special needs.

There has never been a more challenging environment for American manufacturing.

Chapter 11

Dominant Manpower Issues: Education or Extinction

The fundamental question human resources professionals must answer is, "Will there be enough people available, with adequate and appropriate skills, to staff our manufacturing operations?" Answering this question involves the issues of demographics and education.

Workforce demographics will continue to play a dominant role in American manufacturing as we enter the twenty-first century (see Exhibit 11-1). In fact, the importance of demographics for American manufacturing success is likely to continue to grow for several reasons.

As mentioned in Chapter 10, an increasing percentage of the labor force will be women, perhaps 50 percent by the year 2000. This will continue to foster the burgeoning demand for both outside and on-site child care facilities and services. It will also probably mean that the hours worked per employee per week will need to diminish, and that those hours worked need to be as flexible as possible. Women's salaries need to catch up to those of their male counterparts, and in fact are already headed in that direction (although not nearly quickly enough).

Another increasing percentage of the American labor force is minorities, primarily black and Hispanic workers. As a result, education and training, which has been disproportionately lacking in these groups, must be enhanced. As technology content levels increase within manufacturing jobs, it will become increasingly incumbent upon manufacturing companies to sponsor appropriate education and training.

The factor that is most likely to slow this progress in worker education and training is the declining base of American manufacturing jobs. The greatest gains in job availability in America over the next decade will almost certainly be in the service sector. This, combined with the declines projected in the manufacturing base within the United States, will likely mean that manufacturing companies will be in a position to be very selective about the qualifications of potential new employees. It is probable that in an effort to hold the line on costs, American manufacturing companies will be reluctant to make the investments required to develop a world-class workforce.

Faced with the training required to develop and maintain a "world-class" workforce, one manufacturing executive recently told me he preferred to "upgrade the herd" by "swapping out" unskilled workers for more thoroughly trained replacements from outside the company. This kind of quick-fix mentality is increasingly common, particularly among younger executives. It reflects the desire to maximize near-term results while minimizing investment, and gives no quarter

Exhibit 11-1. Dominant manpower issues.

U.S. Manufacturing Manpower					
Business Administration	**Obtaining Business**	**Define Products and Processes**	**Production**	**Distribution**	**Support**
• Aging work force • Growing cultural diversity • Growing employee base in relation to available jobs • Inadequate skills and education • Larger percentage of executives from engineering, information technology, process engineering, international marketing, international supplier management	• Multilingual executives • Cross-functional training in distribution • Design for logistics (DFL) expertise	• Higher percentage of total manufacturing jobs • Smaller/lighter products • Concurrent product/process design • Heavier involvement with customers • Increased emphasis on producibility/distributability	• Technicians, not laborers • Will require extensive education/training • Team-based work • Primary objective is to maintain process controls • Less labor, fewer laborers required • More intimate involvement w/design team	• Fewer people • Merger of functions with marketing organization • Globalized focus • Much more computer- and telecommunications-literate • More outsourced distribution – supplier management skills required	• Extreme data intensity • Heightened levels of computer literacy • High-tech diagnostic repair equipment skills • Desktop publishing/multimedia-based service and tech pubs

to the existing employee base. The concept of mutual loyalty between the company and its employees is evaporating, and will likely continue to do so.

The Changing Performance Yardstick

Contradicting this quick-fix approach, however, are the trends toward decentralized decision making, reductions in layers of management, and the prodigious increase in communications volumes and effectiveness between "mahogany row" and the factory floor. As more decision making is pushed downward, and as organizational lines become more blurred, employees need more training and cross-training. They also need more security than ever, in order to make tough decisions and take risks. The wisest managers will recognize this aspect, and make the investments required to stabilize and enhance existing workforces as much as practical.

Because of shifts in the values of American workers, changes are likely to be made in the ways we provide incentives for, evaluate, and reward our employees. Traditional performance measures in American manufacturing companies have typically centered around direct labor efficiency. At the company level, the measure was most often return on investment (ROI), measured monthly and reported quarterly.

As a rule, this has not been an effective measure because it reflects an end financial result rather than providing a tool for improvement. It is simply too far removed from the actual processes that make the difference between success and failure. As we move through the 1990s, measures for company performance are changing in America.

Unfortunately, that change is occurring very slowly and is likely to continue to be slow. Many corporate executives consider the shift from ROI to RONA (return on net assets) to be a major improvement. It isn't. American manufacturing companies must measure their performance based upon market share. To gain market share, quality must improve (as measured by customer perception) and cycle times must come down. Making these things happen takes patience, time, strategically sound investment, and commitment to a clear company vision. It is unclear when this will be permitted to occur in the United States, because the company shareholders are usually outside investors interested primarily in relatively short-term ROI, and are not particularly interested in market share or long-term growth. It will eventually occur, though, because it is the major deterrent to American competitiveness. It's a case of "change or die."

Objective measures are likely to remain dominant in the 1990s and even the first decade of the next century in the areas of delivery performance (for example, percentage of scheduled deliveries made by schedule date), direct labor efficiency (such as man hours expended per product produced), cost of goods sold (COGS), and defect rates (for example, defects per unit, defects per million). But changes in the values of both workers and customers are likely to cause other measures to be adopted in the areas of process cost, throughput time, customer satisfaction ratings, and total product life cycle profitability.

Whatever areas are measured, the measurement approaches and the incentive and reward systems tied to them are increasingly likely to be tied to processes and to groups (or teams) of people. As processes become more clearly defined for new manufacturing and materials technologies, the natural work groups associated with those processes will be the most effective entities to offer incentives, measure, and reward.

Manpower Used in Business Administration

Manpower issues related to business administration include dealing with a workforce that is aging and becoming increasingly diverse. Extremely skilled leaders will also be increasingly mobile. However, the management and general administration areas will find over the next twenty years that there is a large base of employees and former (usually displaced) employees from which to draw. As a result, manufacturers will become more and more demanding in terms of the depth of education and skills required for employment. In addition, the flattening of organizational structures will require a broader, more multidisciplined knowledge base than ever before. A bachelor-level degree is already almost a prerequisite for management positions in mainstream U.S. manufacturing companies. Advanced degrees such as master's-level degrees will likely replace the B.A. or B.S. as an entry-level standard by the year 2000.

Given the advances forthcoming in materials science and engineering, process engineering, and information technology, it seems reasonable to anticipate that more executives will have successful backgrounds in these disciplines. Considering our propensity to focus on product rather than process engineering and development, it also seems apparent that executives will continue to come from product engineering backgrounds, although this is likely to decline. In many companies

The manpower used in business administration.

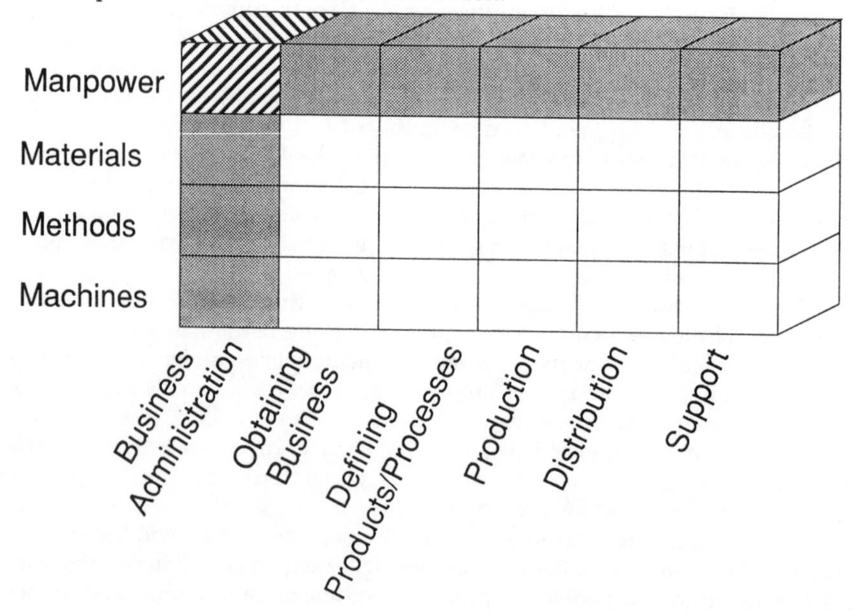

where executives were previously successful emerging from the financial realm, future executives are most likely to come from areas of the manufacturing business pertaining to international marketing or supplier management.

When we review the specific job titles typically comprising the business administration area in American manufacturing firms, and how the pay and training required are likely to evolve over the next ten to twenty years, it seems reasonable to expect a picture like the one depicted in Exhibits 11-2 and 11-3.

Beyond these discrete changes, success in business administration positions will more often depend upon the ability to manage interdisciplinary internal teams, the ability to think in abstract terms, the ability to accurately develop and function in simulation (virtual) environments, and the ability to form and lead international teams.

In addition, one of the most critical success factors for business administration in twenty-first century manufacturing will be the ability to successfully manage the convergence and incorporation of new technologies into the manufacturing process. "Technology management" may well be the hottest curriculum in business schools over the next decade.

It requires a special set of skills (including interpersonal skills) to manage technical efforts in a manner that is strategically wise in view of total-company objectives. For example, the best of our Japanese competitors establish a "sunset" date on every new product line they launch, driving their R&D people to innovation after innovation by establishing these "due dates" for replacement technologies. In the United States, we generally watch for the product line to begin its "mature" phase of the product life cycle, then do whatever we can to "milk" the "cash cow." Replacement technologies are almost as often happenstance as they are structured R&D efforts reaching a planned culmination. Insightful and strategic technology management, driven through interdisciplinary technical development groups performing concurrent product and process development, will be the attribute most sought after in business administration professionals.

The Manpower Used in Obtaining Business

Manpower issues related to obtaining business will be heavily affected over the next two decades by globalized markets and suppliers. The globalization of mainstream businesses will almost certainly mean that multilingual executives will be especially effective in these arenas, particularly over the next ten years. Beyond this point, the incredible capabilities of burgeoning information technologies are likely to significantly reduce (if not nullify) this advantage. However, the ability of an executive to feel comfortable in other cultures and travel extensively and internationally will almost certainly continue to be requisite in the marketing and distribution fields.

Truly successful sales generally occur when one person looks another person in the eye and establishes an agreement—the actual terms and conditions documented later are almost ancillary in most cases. The human contact required to establish adequate trust for sales to occur will not easily be accomplished via electronic communication media. Therefore, the top salespeople will continue to be those who possess the qualities of a good salesperson today. However, the level of work required to support the activities of the salesperson in drafting contract

Exhibit 11-2. Range of wages for business administration jobs in typical mainstream American manufacturing companies.

BUSINESS ADMINISTRATION

JOBS IN TYPICAL MAINSTREAM AMERICAN MANUFACTURING FIRMS

TYPICAL JOB TITLE	1990 TYPICAL RANGE OF WAGES ($000 PER YEAR)		2000 TYPICAL RANGE OF WAGES ($000 PER YEAR)	
	LOW	HIGH	LOW	HIGH
PRESIDENT	150	250	225	370
EXEC. V.P.	125	200	170	256
PROJECT/PROGRAM ADMINISTRATOR	50	75	81	119
ADMIN. ANALYST	20	50	25	63
ADMIN. ASSISTANT	15	35	20	43
OFFICE ASSISTANT	15	35	18	39
CHIEF COUNSEL	100	150	150	225
ATTORNEY	50	100	75	150
AUDITOR	20	60	30	90
PROCEDURES ANALYST	20	50	32	80
V.P. FINANCE	100	175	163	281
CONTROLLER	75	150	119	244
MANAGER - FINANCE	45	70	70	108
ACCOUNTANT	20	60	33	98
CONTRACT ADMINISTRATOR	40	80	68	136
CONTRACT ANALYST	25	70	46	119
FINANCIAL ANALYST	25	70	44	114
COST & ESTIMATING ANALYST	25	70	44	114
V.P. HUMAN RESOURCES	70	125	110	185
MANAGER - H.R.	50	90	70	117
SUPERVISOR - SECURITY	50	80	60	110
H.R. REPRESENTATIVE	25	65	37	89
H.R. ASSISTANT	20	35	25	43
SECURITY OFFICER	15	35	23	46
V.P. INFO. TECHNOLOGY	100	175	165	286
MANAGER - I. T.	50	85	83	124
SYSTEMS ANALYST	20	70	33	116
COMPUTER OPERATOR	15	40	25	60
DATA ENTRY OPERATOR	10	30	15	45
TELECOM. TECHNICIAN	20	35	33	55
COMMUNICATIONS EQUIP. OPERATOR	10	25	15	40

Exhibit 11-3. Educational requirements for business administration jobs in typical mainstream American manufacturing companies.

BUSINESS ADMINISTRATION
JOBS IN TYPICAL MAINSTREAM AMERICAN MANUFACTURING FIRMS

TYPICAL JOB TITLE	EDUCATION TYPICALLY REQUIRED IN 1990	EDUCATION TYPICALLY REQUIRED IN 2000
PRESIDENT	MBA OR EQUIVALENT ADVANCED DEGREE	MBA OR EQUIVALENT ADVANCED DEGREE
EXEC. VICE-PRESIDENT	MBA OR EQUIVALENT ADVANCED DEGREE	MBA OR EQUIVALENT ADVANCED DEGREE
PROJECT/PROGRAM ADMINISTRATOR	B.A. OR B.S. (ALMOST ANY FIELD)	MBA OR EQUIVALENT ADVANCED DEGREE
ADMINISTRATIVE ANALYST	B.A. OR B.S. (ALMOST ANY FIELD)	B.A./B.S. - FINANCE OR BUSINESS
ADMINISTRATIVE ASSISTANT	HIGH SCHOOL DIPLOMA, VOC. TRAIN.	B.A./B.S. - FINANCE OR BUSINESS
OFFICE ASSISTANT	HIGH SCHOOL DIPLOMA, VOC. TRAIN.	ASSOC. DEGREE - BUSINESS, D.P.
CHIEF COUNSEL	JURIS DOCTORATE (J.D.)	JURIS DOCTORATE (J.D.)
ATTORNEY	JURIS DOCTORATE (J.D.)	JURIS DOCTORATE (J.D.)
AUDITOR	B.A. OR B.S. (ALMOST ANY FIELD)	B.A. OR B.S. (ALMOST ANY FIELD)
PROCEDURES ANALYST	B.A. OR B.S. (ALMOST ANY FIELD)	B.A. OR B.S. (ALMOST ANY FIELD)
V.P. FINANCE	MBA, USUALLY A C.P.A.	MBA, USUALLY A C.P.A.
CONTROLLER	MBA, USUALLY A C.P.A.	MBA, USUALLY A C.P.A.
MANAGER - FINANCE	B.A./B.S. - FINANCE OR BUSINESS	MBA, USUALLY A C.P.A.
ACCOUNTANT	B.A. OR B.S. (ALMOST ANY FIELD)	MBA, USUALLY A C.P.A.
CONTRACT ADMINISTRATOR	B.A./B.S. - FINANCE OR BUSINESS	MBA, USUALLY A C.P.A.
CONTRACT ANALYST	B.A./B.S. - FINANCE OR BUSINESS	MBA, USUALLY A C.P.A.
FINANCIAL ANALYST	B.A./B.S. - FINANCE OR BUSINESS	MBA, USUALLY A C.P.A.
COST & ESTIMATING ANALYST	B.A. OR B.S. (ALMOST ANY FIELD)	B.A./B.S. - FINANCE OR BUSINESS
V.P. HUMAN RESOURCES	B.A./B.S.-IND. REL. OR BUSINESS	B.A./B.S.-IND. REL. OR BUSINESS
MANAGER - H.R.	B.A. OR B.S. (ALMOST ANY FIELD)	B.A./B.S.-IND. REL. OR BUSINESS
SUPERVISOR - SECURITY	B.A. OR B.S. (ALMOST ANY FIELD)	B.A. OR B.S.-LAW ENF., BUS. LAW
H.R. REPRESENTATIVE	B.A. OR B.S. (ALMOST ANY FIELD)	B.A. OR B.S. (ALMOST ANY FIELD)
H.R. ASSISTANT	HIGH SCHOOL DIPLOMA	ASSOC. DEGREE-BUS. OR IND. REL.
SECURITY OFFICER	HIGH SCHOOL DIPLOMA	ASSOC. DEGREE-LAW ENF., BUS. LAW
V.P. INFO. TECHNOLOGY	B.A./B.S. - FINANCE OR BUSINESS	B.A./B.S. COMPUTER SCIENCE OR MBA
MANAGER - I.T.	B.A. OR B.S. (ALMOST ANY FIELD)	B.A./B.S. COMP. SCIENCE, LIB. SCI.
SYSTEMS ANALYST	B.A. OR B.S. (ALMOST ANY FIELD)	B.A./B.S. COMPUTER SCIENCE
COMPUTER OPERATOR	HIGH SCHOOL W/VOC. TRAINING	ASSOC. DEGREE - COMPUTER SCIENCE
DATA ENTRY OPERATOR	HIGH SCHOOL W/VOC. TRAINING	HIGH SCHOOL W/VOC. TRAINING
TELECOM. TECHNICIAN	HIGH SCHOOL DIPLOMA	ASSOC. DEGREE - COMPUTER SCIENCE
COMMUNICATIONS EQUIP. OPER.	HIGH SCHOOL DIPLOMA	ASSOC. DEGREE - COMP. SCI. OR ELEC.

language with distributors and customers should be reduced by about half over the next ten years. Machinery like faxes, cellular telephones, and modems will be responsible for much of this improvement.

The education and training of marketing professionals will evolve in many companies over the next decade to include the functions of distribution management as well as sales. It is becoming increasingly clear that marketing and distribution will be performed most effectively when they are combined. The data generated from each discipline's activity is ideal for supporting the other one. For example, historical distribution activity often points to trends that expose fertile areas for marketing efforts related to such areas as after-sales service, spares/repair parts sales, and replacement product sales.

In addition, the people most capable of providing guidance to design engineers regarding product configurations that support rapid and safe delivery of the

The manpower used in obtaining business.

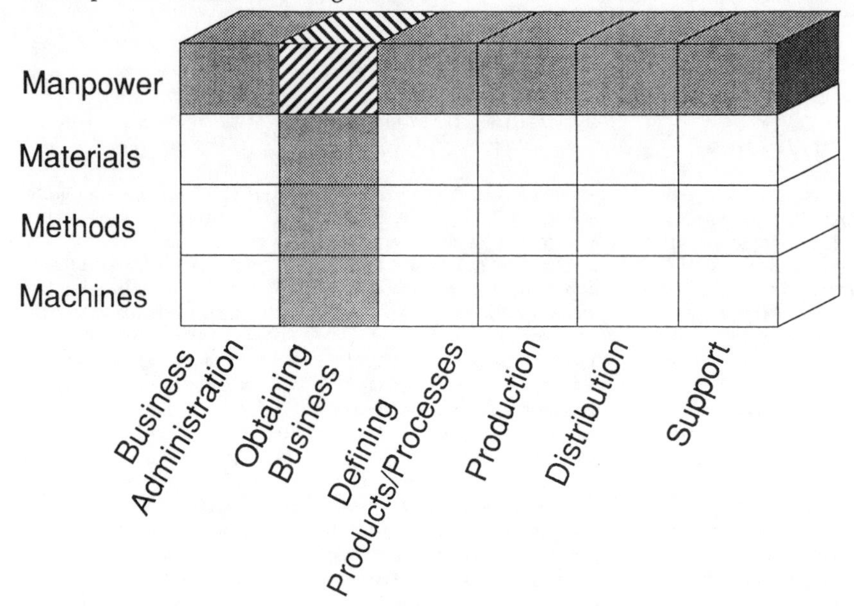

products to their end customers, an approach now being called design for logistics (DFL), are the distribution professionals. Again, this is valuable information when conveyed to and understood by the marketing people. Therefore, some fundamental design knowledge will prove to be useful as well.

An adjunct to this approach is the outsourcing of the bulk of distribution work to companies like Federal Express's Business Logistic Services (BLS). The synergistic value of combining the marketing functions with remaining distribution-related activity in this setting is still retained, but the information and communication linkages between the manufacturer and its distribution service supplier will become extremely important.

When we review the specific jobs typically comprising the "obtaining business" area in American manufacturing firms, pay and training requirements will look something like the picture presented in Exhibits 11-4 and 11-5.

The Manpower Used in Defining Products/Processes

Manpower-related changes in the area of product and process definition may be the most dramatic of all those in the manufacturing disciplines. Because of the sweeping changes in materials and design technologies, both fundamental and ancillary skills will be substantially different. In addition, the percentage of manufacturing jobs comprised of product and process definition professionals will grow significantly. Finally, the tremendous momentum of efforts to reduce cycle-to-market will support continued movement toward concurrency between product/process definition and most of the other primary functions of American manufacturing organizations.

Exhibit 11-4. Range of wages for non-business administration jobs in typical mainstream American manufacturing companies.

OBTAINING BUSINESS

JOBS IN TYPICAL MAINSTREAM AMERICAN MANUFACTURING FIRMS

TYPICAL JOB TITLE	1990 TYPICAL RANGE OF WAGES ($000 PER YEAR)		2000 TYPICAL RANGE OF WAGES ($000 PER YEAR)	
	LOW	HIGH	LOW	HIGH
V.P. MARKETING/BUS. DEV.	100	175	165	288
MANAGER - MARKETING	55	90	88	134
MARKETING REPRESENTATIVE	35	80	55	132
MARKET RESEARCH ANALYST	20	70	33	116
MANAGER - PUBLIC REL.	55	90	80	124
PUBLIC RELATIONS REP.	20	60	30	105
MANAGER - GRAPHICS/PUBS.	40	70	50	88
TECHNICAL WRITER	20	45	28	60
SUPERVISOR - GRAPHICS	30	55	38	68
ARTIST/ILLUSTRATOR	20	50	28	69
GRAPHICS/PUBLICATION TECH.	20	45	25	55

Exhibit 11-5. Educational requirements for non-business administration jobs in typical mainstream American manufacturing firms.

OBTAINING BUSINESS

JOBS IN TYPICAL MAINSTREAM AMERICAN MANUFACTURING FIRMS

TYPICAL JOB TITLE	EDUCATION TYPICALLY REQUIRED IN 1990	EDUCATION TYPICALLY REQUIRED IN 2000
V.P. MARKETING/BUS. DEV.	B.A./B.S.-BUSINESS OR FINANCE	MBA OR EQUIVALENT ADVANCED DEGREE
MANAGER - MARKETING	B.A. OR B.S. (ALMOST ANY FIELD)	B.A./B.S.-BUSINESS, MARKETING OR MBA
MARKETING REPRESENTATIVE	B.A. OR B.S. (ALMOST ANY FIELD)	B.A./B.S.-BUSINESS, MARKETING OR MBA
MARKET RESEARCH ANALYST	B.A. OR B.S. (ALMOST ANY FIELD)	B.A./B.S.-BUSINESS, MARKETING OR MBA
MANAGER - PUBLIC RELATIONS	B.A. OR B.S. (ALMOST ANY FIELD)	B.A./B.S.-BUSINESS, MARKETING OR MBA
PUBLIC RELATIONS REP.	B.A. OR B.S. (ALMOST ANY FIELD)	B.A./B.S.-BUSINESS, MARKETING
MANAGER - GRAPHICS/PUBS.	B.A. OR B.S. (ALMOST ANY FIELD)	B.S./B.S. BUS., MARKTG., LIB. SCI.
TECHNICAL WRITER	B.A. OR B.S. (ALMOST ANY FIELD)	B.S./B.S. BUS., MARKTG., LIB. SCI.
SUPERVISOR - GRAPHICS	HIGH SCHOOL, W/VOC. TRAINING	ASSOC. DEGREE - COMPUTER SCIENCE
ARTIST/ILLUSTRATOR	HIGH SCHOOL, W/VOC. TRAINING	HIGH SCHOOL, W/VOC. TRAINING
GRAPHICS/PUBLICATION TECH.	HIGH SCHOOL DIPLOMA	HIGH SCHOOL, W/VOC. TRAINING

The manpower used in defining products/processes.

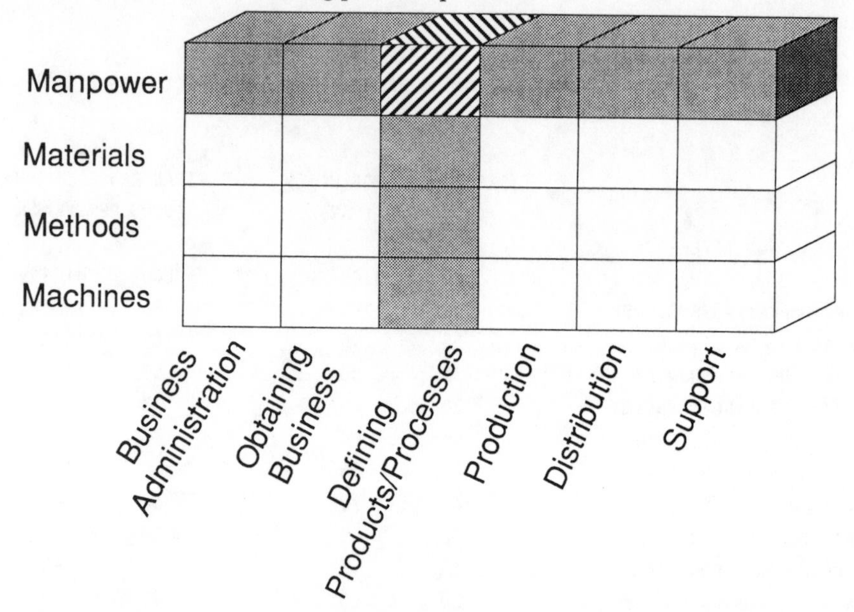

New Materials Require New Know-How

As discussed in Chapter 6, materials science and materials engineering are likely to be extremely powerful forces in shaping the design, composition, and manufacturing processes of our future products. They will influence the designs by allowing products to have fewer individually fabricated components, by allowing the same functions to be performed by smaller and lighter products, and by providing contour and surface capabilities that were impossible to achieve with earlier material compositions. They will influence the composition of the products by reducing weight through improved strength, thermal resistance, electrical conductivity, and other desirable characteristics. The composition of new materials will largely determine the types of manufacturing processes required (for example, composites processing requires autoclaves, plastics use injection molding equipment, metals utilize presses).

Because of these developments, product design will require knowledge of broad ranges of new materials and their properties in order to ensure that the products are able to become and remain competitive. Professionals in this field will also need to know the fundamental processes utilized to produce components from these new materials in order to ensure design producibility and existing manufacturing capability.

Who's Designing Our Future?

Design technologies themselves have seen major changes over the last few decades, as discussed in Chapter 5. We have moved from paper drawings to mylar to 2-D computer-aided design (CAD) drawings to 3-D "solid" models. The advent of virtual reality and other less exotic simulation software and hardware will require different and broader knowledge of design engineers in American manufacturing companies.

Unfortunately, the systems and software utilized by different companies are frequently significantly different. Therefore, it will be difficult to move seamlessly from one company or system to another. Substantial retraining will be required in these situations for at least the next decade, and perhaps for much longer unless a great deal of progress is made toward standardization in this field.

Today there is a trend in manufacturing called concurrency; concurrent engineering is the simultaneous development of products and the processes that produce them. Because of the nature of design technology changes and the movement toward concurrency, the percentage of an organization classified as "product and process definition" will grow significantly. In fact, over the next two decades, company functions associated with product and process definition are likely to double from the current level of 20 to 25 percent to nearly 50 percent in many firms.

The advent of concurrent engineering will mean that design activities will change to include the following:

- A more direct role in identifying and interpreting customer needs. Teams of customers and potential customers will be utilized increasingly as panels of experts to identify desirable product and service features. Product design teams will then employ approaches like quality function deployment (QFD) to develop product characteristics that satisfy those needs.

- Concurrent design team members from the production organization will help the team to ensure that all product designs are producible given existing or planned manufacturing capabilities and ensure that manufacturing capabilities are enhanced as required. They will also assist in defining "critical product characteristics" from the standpoint of fabrication and assembly, so that process controls can be implemented almost immediately on critical manufacturing processes. In addition, they will assist in identifying supplier packaging and delivery requirements for those components not fabricated in-house. Often, they will perform rough-cut capacity analysis and planning as the process designs are completed, especially on potential "bottleneck" operations and processes to ensure adequate manufacturing capacity through development and early stages of production. Finally, these team members will use (still in development) interface software to write the programs that operate fabrication equipment to produce the parts.

- Concurrent design team members who specialize in distribution will ensure that all anticipated logistics requirements are incorporated into the design of the product and processes, such that postproduction handling is minimized, shelf life considerations are recognized and appropriately dealt with, and "distribution friendly" packaging is incorporated.

- Concurrent design team members who specialize in support of customers and products will be responsible for ensuring that supportability considerations are built into the products. They will utilize various product- and service-testing approaches to ensure that product failures and service requirements are anticipated, identified, and provided for prior to new product launch. Increasingly, simulation skills and modeling/simulation tools will become important parts of this work.

In addition, it will be increasingly popular to have the process experts who have developed the manufacturing, distribution, and support aspects of the (concurrent) design move to the manufacturing, distribution, and service organizations with the new product. In this manner, they will live with the successes and failures of their work, and the lessons learned from this experience will enrich the entire organization.

A look at the pay and training requirements for specific job titles typically comprising the "defining products and processes" area in American manufacturing firms shows that the picture looks brighter for the people skilled in this area than for any of the other manufacturing disciplines (see Exhibits 11-6 and 11-7).

The Manpower Used in Production

Manpower-related changes pertaining to the production area of manufacturing organizations over the next two decades will center around a metamorphosis in skill sets and a decline in manpower levels.

The age of livable wages for low-skill jobs is nearly over. Entry-level laborer

Exhibit 11-6. Range of wages for product and process definition jobs in typical mainstream American manufacturing companies.

PRODUCT & PROCESS DEFINITION
JOBS IN TYPICAL MAINSTREAM AMERICAN MANUFACTURING FIRMS

TYPICAL JOB TITLE	1990 TYPICAL RANGE OF WAGES ($000 PER YEAR)		2000 TYPICAL RANGE OF WAGES ($000 PER YEAR)	
	LOW	HIGH	LOW	HIGH
V.P. ENGINEERING	100	175	175	303
MANAGER - PROD. ENGINEERING	50	100	83	149
SENIOR ENGINEER	40	80	70	125
DESIGN ENGINEER	25	70	44	120
DRAFTSMAN	20	40	23	45
CONFIGURATION MANAGER	40	70	56	91
CONFIGURATION ANALYST	20	55	28	75
MANAGER - R&D	50	90	88	125
R&D ENGINEER	25	70	48	123
R&D TECHNICIAN	20	50	33	81
MANAGER - MANUF. ENG.	50	80	75	110
MANUFACTURING ENGINEER	20	60	22	96

Exhibit 11-7. Educational requirements for product and process definition jobs in typical mainstream American manufacturing companies.

PRODUCT & PROCESS DEFINITION
JOBS IN TYPICAL MAINSTREAM AMERICAN MANUFACTURING FIRMS

TYPICAL JOB TITLE	EDUCATION TYPICALLY REQUIRED IN 1990	EDUCATION TYPICALLY REQUIRED IN 2000
V.P. ENGINEERING	MASTERS LEVEL ENGINEERING	PH.D.-ENG., MATERIALS, CHEMISTRY
MANAGER-PROD. ENGINEERING	BACHELORS LEVEL ENGINEERING	MASTERS LEVEL ENGINEERING
SENIOR ENGINEER	BACHELORS LEVEL ENGINEERING	MASTERS LEVEL ENGINEERING
DESIGN ENGINEER	BACHELORS LEVEL ENGINEERING	BACHELORS LEVEL ENGINEERING
DRAFTSMAN	HIGH SCHOOL W/VOC. TRAINING	BACHELORS LEVEL ENGINEERING
CONFIGURATION MANAGER	B.A. OR B.S. (ALMOST ANY FIELD)	MASTERS LEVEL ENGINEERING
CONFIGURATION ANALYST	B.A. OR B.S. (ALMOST ANY FIELD)	BACHELORS LEVEL ENGINEERING
MANAGER - R&D	BACHELORS LEVEL ENGINEERING	PH.D.-ENG., MATERIALS, CHEM. ELEC
R&D ENGINEER	BACHELORS LEVEL ENGINEERING	MASTERS LEVEL-ENG, MATER, CHEM, ELEC
R&D TECHNICIAN	HIGH SCHOOL DIPLOMA	B.A./B.S.- ENG, CHEM, ELEC, MATER.
MANAGER - MANUF. ENG.	BACHELORS LEVEL ENGINEERING	MASTERS LEVEL ENGINEERING
MANUFACTURING ENGINEER	B.A. OR B.S. (ALMOST ANY FIELD)	BACHELORS LEVEL ENGINEERING

The manpower used in production.

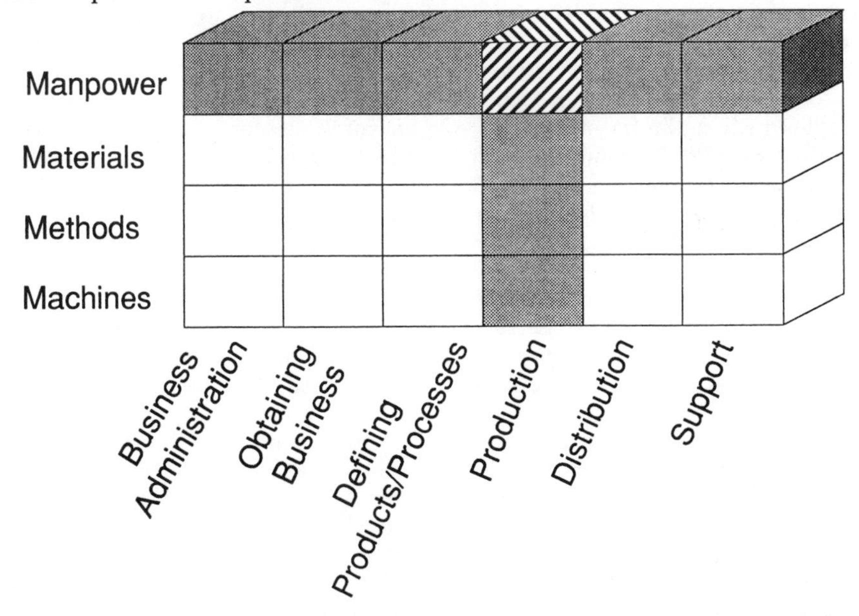

jobs for high school graduates, abundant through the 1960s and 1970s, are all but gone in the United States today. Certainly, the potential for a successful career in mainstream American manufacturing with only a high school diploma is gone. Fabrication and assembly technicians over the next two decades will require skills in numerical control (NC) programming, formal problem solving, statistical process control, design of experiments, and value analysis/value engineering. They will be required to operate lasers and water jets more often than jigsaws and shears. They will work with temperatures, pressures, and cure times more often than punches, presses, and bend-brakes. They will work in teams that perform processes rather than work as individuals performing discrete fabrication operations. Methods like electron beam (EB) welding, superplastic forming, and electrochemical machining will be far more commonplace than many of today's dated techniques.

From the Parts to the Whole

The skills required to perform these operations are somewhat specialized, to be sure. However, more important than the specialized, process-specific skills are those skills that allow the factory worker of the future to manage entire processes for families of parts, keep the processes within (statistical) control, and anticipate and prevent defects. These fundamental skills involve training and education that are not yet significant parts of our formal education curricula. As a result, today's workers are picking them up very slowly from on-the-job training and company-sponsored seminars. Generally, the efforts are very focused, narrowly applied to specific individuals or processes, and not well retained.

The advent of automation and other new manufacturing technologies, coupled with advances in information technology and new materials developments, will serve to dramatically reduce the quantity of general factory floor labor required. This poses interesting problems for shop floor supervision as well as significant opportunities.

Factory floor supervision is a job classification that can generally be considered an "endangered species." In fact, the number of supervisors required per unit produced will decline dramatically by the turn of the century. The primary reason for this is that the amount of direct labor required per unit is going to decline dramatically. Automation and the drive to out-source increasing percentages of our components, especially to foreign suppliers, are largely responsible.

In addition, the skills required for shop floor supervision will change. Future front-line factory management will require facilitating and coaching skills, because the direct labor employees will increasingly be organized into "teams" mirroring the group technology application of manufacturing cells and flexible manufacturing systems. It will not be enough to be the best machine operator in the group to qualify as a supervisor. Future supervisors will need to be leaders, not merely technically or physically skilled workers.

Generally, direct labor productivity will greatly improve as fewer people are required to generate the same number of products. However, the factory promises to become a more "capital intense" environment, which will hold interesting ramifications for the return on net assets (RONA) calculations used to measure company performance.

A picture of the salaries and educational requirements for specific job titles typically comprising the "production" area in American manufacturing firms, and how they are likely to evolve over the next ten to twenty years, is shown in Exhibits 11-8 and 11-9. To quote a popular automotive advertisement, "The rules have changed."

The Manpower Used in Distribution

Manpower-related changes associated with the distribution function of American manufacturing businesses will involve reductions in overall quantities of people, a

(*Text continues on page 144.*)

Exhibit 11-8. Range of wages for production jobs in typical mainstream American manufacturing companies.

PRODUCTION

JOBS IN TYPICAL MAINSTREAM AMERICAN MANUFACTURING FIRMS

TYPICAL JOB TITLE	1990 TYPICAL RANGE OF WAGES ($000 PER YEAR)		2000 TYPICAL RANGE OF WAGES ($000 PER YEAR)	
	LOW	HIGH	LOW	HIGH
V. P. OPERATIONS	100	160	150	240
DIRECTOR - MANUFACTURING	75	125	101	170
MANAGER - FABRICATION OPS.	50	100	69	128
FABRICATION - SUPERVISOR	30	75	40	92
FABRICATION TECHNICIAN	15	40	28	66
ELECTRONICS TECHNICIAN	15	40	21	72
CHEMICAL PROCESSOR	15	35	21	59
MACHINE OPERATOR	15	35	20	43
FINISHER	15	35	20	43
TOOLING SUPERVISOR	35	60	46	83
FABRICATION TOOL MAKER	15	40	23	55
FABRICATION LABORER	15	30	18	34
TEST SUPERVISOR	20	45	30	65
TEST TECHNICIAN	15	35	25	50
MANAGER-ASSEMBLY OPERATIONS	50	100	70	130
ASSEMBLY SUPERVISOR	30	75	41	95
ASSEMBLER	15	40	21	52
ELECTRICIAN	15	40	25	60
WELDER	25	35	34	44
PAINTER	15	35	23	46
ASSEMBLY LABORER	15	30	20	38
ASSEMBLY TOOL MAKER	15	40	23	56
TEST SUPERVISOR	20	45	30	60
TEST TECHNICIAN	15	35	25	50
INDUSTRIAL TRUCK OPERATOR	15	30	20	38
DIRECTOR - MATERIALS	75	125	119	181
MANAGER-PROD. & INVENT. MGMT.	35	55	50	74
MASTER SCHEDULER	25	75	43	117
SHOP FLOOR SCHEDULER	25	40	36	55
EXPEDITER	15	30	18	34
INVENTORY MGMT SUPERVISOR	30	45	34	49
INVENTORY ANALYST	25	40	33	50
PRODUCTION PLANNER	20	45	28	60
CYCLE COUNTER	15	30	20	38
MANAGER-RCVG/SHIP/STORE/TRANSP	35	60	55	89

(*continues*)

Exhibit 11-8. (continued)

PRODUCTION, PAGE 2

JOBS IN TYPICAL MAINSTREAM AMERICAN MANUFACTURING FIRMS

TYPICAL JOB TITLE	1990 TYPICAL RANGE OF WAGES ($000 PER YEAR)		2000 TYPICAL RANGE OF WAGES ($000 PER YEAR)	
	LOW	HIGH	LOW	HIGH
RECEIVING SUPERVISOR	30	45	45	60
MATERIAL HANDLER	10	30	13	38
MATERIAL CONTROL CLERK	15	30	20	38
SHIPPING SUPERVISOR	30	45	45	60
PACKER/SHIPPER	15	30	23	41
SHIPPING CLERK	15	30	23	41
DELIVERY DRIVER	25	35	37	47
STORES SUPERVISOR	30	45	41	60
STORES OPERATOR	15	30	20	38
WAREHOUSE COORDINATOR	25	40	33	50
TRANSPORTATION SUPERVISOR	35	50	50	75
TRANS./TRAFFIC ANALYST	25	40	40	60
MANAGER - PROCUREMENT	45	95	75	146
PROCUREMENT ADMINISTRATOR	40	75	56	96
SUBCONTRACTS ADMINISTRATOR	30	50	50	74
PURCHASING AGENT	40	55	66	88
BUYER	20	45	33	71
MANAGER-INDUST. ENGINEER	45	75	65	101
INDUSTRIAL ENGINEER	20	50	30	75
MANAGER - FACILITIES & MAINT.	45	75	65	101
FACILITIES ENGINEER	25	50	40	75
FACILITIES TECHNICIAN	15	35	23	46
MAINTENANCE SUPERVISOR	30	45	40	58
MAINTENANCE TECHNICIAN	10	30	15	45
MANAGER - QUALITY ASSURANCE	30	75	49	108
QUALITY ENGINEER	20	60	33	98
METROLOGY SUPERVISOR	25	45	40	64
METROLOGY TECHNICIAN	15	30	28	49
QUALITY CONTROL ANALYST	15	30	28	49
INSPECTION SUPERVISOR	30	45	45	64
INSPECTOR	15	40	23	55
MGR - ENVIRON, HEALTH & SAFETY	30	75	50	109
E H & S ADMINISTRATOR	25	50	40	75
E H & S ENGINEER	20	40	35	70

Exhibit 11-9. Educational requirements for production jobs in typical mainstream American manufacturing companies.

PRODUCTION

JOBS IN TYPICAL MAINSTREAM AMERICAN MANUFACTURING FIRMS

TYPICAL JOB TITLE	EDUCATION TYPICALLY REQUIRED IN 1990	EDUCATION TYPICALLY REQUIRED IN 2000
V.P. OPERATIONS	B.A. OR B.S. (ALMOST ANY FIELD)	MBA, OR MASTERS LEVEL IND. MGMT.
DIRECTOR - MANUFACTURING	B.A. OR B.S. (ALMOST ANY FIELD)	MBA, OR MASTERS LEVEL IND. MGMT.
MANAGER - FABRICATION OPS.	B.A. OR B.S. (ALMOST ANY FIELD)	B.A./B.S.-BUS. OR IND. MGMT.
FABRICATION SUPERVISOR	HIGH SCHOOL W/VOC. TRAINING	ASSOC OR B.S.-MECH/ELEC/CHEM ENG.
FABRICATION TECHNICIAN	HIGH SCHOOL DIPLOMA	B.S.-MECH/ELEC/CHEM ENG.
ELECTRONICS TECHNICIAN	HIGH SCHOOL W/VOC. TRAINING	ASSOC LEVEL - ELEC ENG.
CHEMICAL PROCESSOR	HIGH SCHOOL W/VOC. TRAINING	B.S. - CHEM ENGINEERING
MACHINE OPERATOR	HIGH SCHOOL DIPLOMA	ASSOC. DEGREE - MECH ENGINEERING
FINISHER	HIGH SCHOOL DIPLOMA	HIGH SCHOOL W/VOC. TRAINING
TOOLING SUPERVISOR	HIGH SCHOOL W/VOC. TRAINING	ASSOC. DEGREE - MECH ENGINEERING
FAB. TOOL MAKER	HIGH SCHOOL W/VOC. TRAINING	HIGH SCHOOL W/VOC. TRAINING
FABRICATION LABORER	HIGH SCHOOL DIPLOMA	HIGH SCHOOL W/VOC. TRAINING
TEST SUPERVISOR	ASSOC DEGREE OR HIGH SCHOOL W/VOC.	B.S.-ELEC OR CHEM. ENG.
TEST TECHNICIAN	HIGH SCHOOL DIPLOMA W/VOC.	ASSOC LEVEL - MECH/ELEC/CHEM. ENG.
MANAGER - ASSEMBLY OPS.	B.A. OR B.S. (ALMOST ANY FIELD)	B.S. - MECH/ELEC/CHEM ENG.
ASSEMBLY SUPERVISOR	HIGH SCHOOL DIPLOMA	ASSOC DEGREE - MECH/ELEC/CHEM ENG.
ASSEMBLER	HIGH SCHOOL DIPLOMA	ASSOC DEGREE OR HIGH SCHOOL W/VOC.
ELECTRICIAN	HIGH SCHOOL W/VOC. TRAINING	ASSOC DEGREE - ELEC. OR MECH. ENG.
WELDER	HIGH SCHOOL DIPLOMA	HIGH SCHOOL W/VOC. TRAINING
PAINTER	HIGH SCHOOL DIPLOMA	HIGH SCHOOL W/VOC. TRAINING
ASSEMBLY LABORER	HIGH SCHOOL DIPLOMA	HIGH SCHOOL W/VOC. TRAINING
ASSEMBLY TOOL MAKER	HIGH SCHOOL DIPLOMA	ASSOC DEGREE OR HIGH SCHOOL W/VOC.
TEST SUPERVISOR	B.A. OR B.S. (ALMOST ANY FIELD)	B.S.- MECH ENG. OR ELEC ENG.
TEST TECHNICIAN	ASSOC DEGREE OR HIGH SCHOOL W/VOC.	ASSOC DEGREE - MECH/ELEC ENG.
INDUST. TRUCK OPERATOR	HIGH SCHOOL DIPLOMA	HIGH SCHOOL DIPLOMA
DIRECTOR - MATERIALS	B.A. OR B.S. (ALMOST ANY FIELD)	MASTERS - MAT. SCI. OR MECH. ENG.
MANAGER-PROD. & INVENT. MGMT.	B.A. OR B.S. (ALMOST ANY FIELD)	B.A./B.S.- BUS., IND. MGMT, OR MBA
MASTER SCHEDULER	B.A. OR B.S. (ALMOST ANY FIELD)	B.A./B.S.- BUS., IND. MGMT.
SHOP FLOOR SCHEDULER	ASSOC. DEGREE OR EQUIVALENT	B.A./B.S.- BUS., IND. MGMT.
EXPEDITER	HIGH SCHOOL DIPLOMA	ASSOC. DEGREE OR EQUIVALENT
INVENTORY MGMT. SUPERVISOR	B.A. OR B.S. (ALMOST ANY FIELD)	B.A./B.S.- BUS., IND. MGMT, OR MBA
INVENTORY ANALYST	B.A. OR B.S. (ALMOST ANY FIELD)	B.A./B.S.- BUS., IND. MGMT, OR MBA
PRODUCTION PLANNER	HIGH SCHOOL DIPLOMA	B.A./B.S.- BUS., IND. MGMT.
CYCLE COUNTER	HIGH SCHOOL DIPLOMA	HIGH SCHOOL DIPLOMA
MGR-RCVG/SHIP/STORES/TRANSP.	B.A. OR B.S. (ALMOST ANY FIELD)	B.A./B.S.- BUS., IND. MGMT.
RECEIVING SUPERVISOR	HIGH SCHOOL DIPLOMA	HIGH SCHOOL DIPLOMA
MATERIAL HANDLER	HIGH SCHOOL DIPLOMA	HIGH SCHOOL DIPLOMA
MATERIAL CONTROL CLERK	HIGH SCHOOL DIPLOMA	HIGH SCHOOL DIPLOMA
SHIPPING SUPERVISOR	HIGH SCHOOL DIPLOMA	B.A./B.S.- BUS., IND. MGMT.
PACKER/SHIPPER	HIGH SCHOOL DIPLOMA	HIGH SCHOOL DIPLOMA
SHIPPING CLERK	HIGH SCHOOL DIPLOMA	HIGH SCHOOL DIPLOMA
DELIVERY DRIVER	HIGH SCHOOL DIPLOMA	HIGH SCHOOL DIPLOMA
STORES SUPERVISOR	HIGH SCHOOL DIPLOMA	ASSOC. DEGREE OR EQUIVALENT
STORES OPERATOR	HIGH SCHOOL DIPLOMA	HIGH SCHOOL DIPLOMA
WAREHOUSE COORDINATOR	HIGH SCHOOL DIPLOMA	HIGH SCHOOL DIPLOMA
TRANSPORTATION SUPERVISOR	HIGH SCHOOL DIPLOMA	B.A./B.S. - BUS., IND. MGMT.
TRANS./TRAFFIC ANALYST	B.A. OR B.S. (ALMOST ANY FIELD)	B.A./B.S. - BUS., IND. MGMT.

(*continues*)

Exhibit 11-9. (continued)

PRODUCTION (Page 2)
JOBS IN TYPICAL MAINSTREAM AMERICAN MANUFACTURING FIRMS

TYPICAL JOB TITLE	EDUCATION TYPICALLY REQUIRED IN 1990	EDUCATION TYPICALLY REQUIRED IN 2000
MANAGER - PROCUREMENT	B.A. OR B.S. (ALMOST ANY FIELD)	MBA, OR MASTERS LEVEL IND. MGMT.
PROCUREMENT ADMINISTRATOR	B.A. OR B.S. (ALMOST ANY FIELD)	MBA, OR B.A./B.S. IN BUS, BUS LAW
SUBCONTRACTS ADMINISTRATOR	B.A. OR B.S. (ALMOST ANY FIELD)	MBA, OR B.A./B.S. IN BUS, BUS LAW
PURCHASE AGENT	B.A. OR B.S. (ALMOST ANY FIELD)	B.A./B.S. - BUS., IND. MGMT.
BUYER	B.A. OR B.S. (ALMOST ANY FIELD)	B.A./B.S. - BUS., IND. MGMT.
MANAGER - INDUST. ENGINEER	B.A. OR B.S. (ALMOST ANY FIELD)	MASTERS - IND. MGMT., IND./MECH. ENG.
INDUSTRIAL ENGINEER	B.A. OR B.S. (ALMOST ANY FIELD)	B.A./B.S. - IND. MGMT., IND./MECH. ENG.
MANAGER - FACILITIES & MAINT.	B.A. OR B.S. (ALMOST ANY FIELD)	MASTERS - IND. MGMT., IND./MECH. ENG.
FACILITIES ENGINEER	B.A. OR B.S. (ALMOST ANY FIELD)	B.A./B.S.- IND. MGMT., IND./MECH. ENG.
FACILITIES TECHNICIAN	HIGH SCHOOL DIPLOMA	HIGH SCHOOL DIPLOMA
MAINTENANCE SUPERVISOR	B.A. OR B.S. OR ASSOC./VOC.	B.A./B.S. - MECH ENG.
MAINTENANCE TECHNICIAN	HIGH SCHOOL W/VOC. TRAINING	ASSOC. DEGREE - MECH. ENG.
MANAGER - QUALITY ASSURANCE	B.A. OR B.S. (ALMOST ANY FIELD)	B.A./B.S. - BUS., IND. MGMT.
QUALITY ENGINEER	B.A. OR B.S. (ALMOST ANY FIELD)	B.A./B.S. - MECH. ENG.
METROLOGY SUPERVISOR	B.A. OR B.S. (ALMOST ANY FIELD)	B.A./B.S. - MECH. ENG.
METROLOGY TECHNICIAN	B.A. OR B.S. (ALMOST ANY FIELD)	ASSOC. DEGREE - MECH. ENG., ELEC.
QUALITY CONTROL ANALYST	B.A. OR B.S. (ALMOST ANY FIELD)	B.A./B.S. - MECH. ENG.
INSPECTION SUPERVISOR	HIGH SCHOOL W/VOC. TRAINING	B.A./B.S. - MECH. ENG.
INSPECTOR	HIGH SCHOOL DIPLOMA	ASSOC. DEGREE - MECH. ENG., ELEC.
MGR.-ENVIRON, HEALTH & SAFETY	B.A. OR B.S. (ALMOST ANY FIELD)	MASTERS - IND. MGMT., MECH ENG.
E H & S ADMINISTRATOR	B.A. OR B.S. (ALMOST ANY FIELD)	B.A./B.S. - IND. MGMT. MECH ENG.
E H & S ENGINEER	BACHELOR LEVEL-ENGINEERING	B.A./B.S. - MECH/ELEC/CHEM ENG.

continuing shift toward information-oriented job functions, a general merger with marketing organizations, and, in many cases, a relegation of actual distribution functions to outside entities. Like the marketing organization, distribution will change significantly in scope and focus as the globalization of markets continues.

Efficiencies gained in transportation, warehousing, packaging, and general material-handling technologies will reduce the number of people required to distribute goods over the next two decades. Bar coding and enhanced material-handling devices such as automated guided vehicles (AGVs), automated storage and retrieval systems (AS/RSs), and other electronic and automation marvels will continue to erode the manual labor hours required per load of goods distributed. Again, this means an improved picture for direct labor efficiency, but a more capital-intense distribution operation.

Other aspects of distribution that are primarily information-based, such as the routing of trucks and deployment of inventory among decentralized distribution centers, are likely to benefit from recent and forthcoming improvements in information and telecommunications technology. Satellite-based navigation and tracking of vehicles, fax machines, cellular telephones, and other telecommunications advances still to come will dramatically reduce the time required for the travel of goods, as well as the often even more prodigious cycle time of paperwork processing associated with the distribution of goods. This will require increasingly computer- and telecommunications-literate distribution professionals.

The manpower used in distribution.

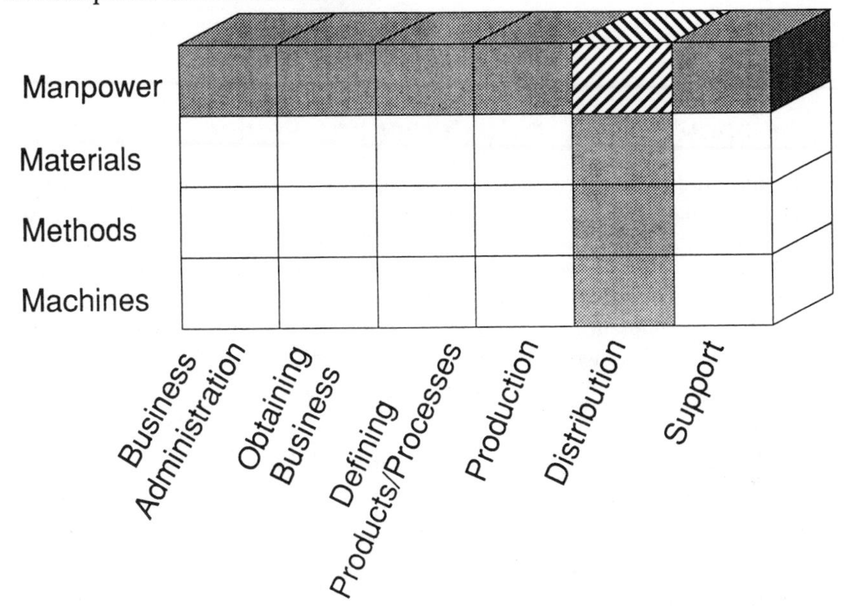

The merger of distribution and marketing functions will of course mean that the knowledge and skill base of people who have been strictly distribution professionals will need to be enhanced to include some knowledge of the marketing process. Although people will still tend to specialize in these different functions, the sheer commonality of information that is the lifeblood of both functions will draw them irresistibly together. Therefore, professionals in the distribution area will need at least enough knowledge of marketing principles to be conversant in the area, and will find specialized knowledge of their own company's marketing practices to be invaluable.

It is fair to say that a thorough understanding of both areas will be an enormous asset to those who plan to reach executive ranks in the new marketing and distribution organizations that develop over the next decade or so in mainstream manufacturing companies. This will mean that a command of distribution requirements planning (DRP) logic and processes will be required, as well as a good understanding of quality function deployment (QFD), customer needs identification, order entry processes, automated market-information-gathering techniques, rapid prototyping concepts, and general sales principles.

Finally, there is likely to be a continued movement on the part of many mainstream American manufacturers toward the use of outside companies to perform their distribution functions. Outside distributors now offer a broad range of warehousing, distribution, delivery, and even quality assurance services that manufacturers are finding to be increasingly attractive alternatives to maintaining large support staffs (see the discussion on distribution methods in Chapter 5).

This means that there will be fewer professional distribution people within the manufacturing organization, but it also offers important opportunities within the service sector for the consummate distribution professional.

Exhibits 11-10 and 11-11 depict the likely scenario for salaries and educational requirements respectively for specific job titles typically part of the "distribution" area in American manufacturing firms as they evolve over the next ten to twenty years.

The Manpower Used in Customer and Product Support

Professionals in the customer and product support area of American manufacturing companies through the first decade of the twenty-first century will still be required to understand the fundamentals of service part sales forecasting, service part delivery, field service repair team training and management, defective product disposition, warranty claim management, production of technical publications, and the collection and utilization of field performance data. However, the skills they will need to do this work will change significantly.

Data Literacy in Demand

Forecasting the sales levels of service parts will involve the use of unprecedented volumes of data, virtually all of which will be electronically accessed

Exhibit 11-10. Range of wages for distribution jobs in typical mainstream American manufacturing companies.

DISTRIBUTION

JOBS IN TYPICAL MAINSTREAM AMERICAN MANUFACTURING FIRMS

TYPICAL JOB TITLE	1990 TYPICAL RANGE OF WAGES ($000 PER YEAR)		2000 TYPICAL RANGE OF WAGES ($000 PER YEAR)	
	LOW	HIGH	LOW	HIGH
V.P. OF DISTRIBUTION	80	140	120	210
MANAGER-DIST. REQMTS. PLANNING	30	75	45	101
DIST. REQMTS. PLANNER	25	55	43	78
LOGISTICS ANALYST	25	45	43	75
GEN. MANAGER-DIST. CENTER	45	100	70	150
DIST. CNTR. RECVG. SUPERVISOR	30	45	42	65
DIST. CNTR. SHIPPING SUPERVISOR	35	55	50	80
DIST. WAREHOUSE SUPERVISOR	30	50	40	66
DIST. MATERIAL HANDLER	10	30	13	38
DIST. RCVG./SHIPPING CLERK	15	30	23	41
DIST. TRANSPORT. SUPERVISOR	25	55	40	80
DIST. TRANSPORT. ANALYST	25	40	40	60
DIST. CNTR. INVEN. MGR.	25	50	40	69
DIST. CNTR. INVEN. ANALYST	15	40	25	60
DIST. CNTR. CYCLE COUNTER	10	25	13	33
DIST. CNTR. DELIV. DRIVER	25	40	40	60

Exhibit 11-11. Educational requirements for distribution jobs in typical mainstream American manufacturing companies.

DISTRIBUTION

JOBS IN TYPICAL MAINSTREAM AMERICAN MANUFACTURING FIRMS

TYPICAL JOB TITLE	EDUCATION TYPICALLY REQUIRED IN 1990	EDUCATION TYPICALLY REQUIRED IN 2000
V.P. OF DISTRIBUTION	B.A./B.S.-BUSINESS OR FINANCE	MBA OR B.A./B.S.-BUS. OR FINANCE
MANAGER-DIST. REQ. PLANNING	B.A. OR B.S. (ALMOST ANY FIELD)	MBA OR B.A./B.S.-BUS. OR FINANCE
DIST. REQUIREMENTS PLANNER	HIGH SCHOOL DIPLOMA	B.A./B.S.-BUSINESS OR FINANCE
LOGISTICS ANALYST	B.A. OR B.S. (ALMOST ANY FIELD)	B.A./B.S.-BUSINESS OR FINANCE
GEN. MGR.-DISTRIBUTION CENTER	B.A. OR B.S. (ALMOST ANY FIELD)	MBA OR B.A./B.S.-IND. MGMT./FINANCE
DIST. CNTR. RECVG. SUPERVISOR	HIGH SCHOOL DIPLOMA	B.A./B.S.- IND. MGMT.
DIST. CNTR.SHIPPING SUPV.	HIGH SCHOOL DIPLOMA	B.A./B.S.- IND. MGMT.
DIST. WAREHOUSE SUPERVISOR	HIGH SCHOOL DIPLOMA	B.A./B.S.- IND. MGMT.
DIST. MATERIAL HANDLER	HIGH SCHOOL DIPLOMA	HIGH SCHOOL DIPLOMA
DIST. RECVG./SHIPPING CLERK	HIGH SCHOOL DIPLOMA	ASSOC. LEVEL-BUS. OR FINANCE
DIST. TRANSPORT. SUPERVISOR	B.A. OR B.S. (ALMOST ANY FIELD)	B.A./B.S.- IND. MGMT.
DIST. TRANSPORT. ANALYST	B.A. OR B.S. (ALMOST ANY FIELD)	B.A./B.S.- IND. MGMT.
DIST. CNTR. INVEN. MGR.	B.A. OR B.S. (ALMOST ANY FIELD)	B.A./B.S.- IND. MGMT.
DIST. CNTR. INVEN. ANALYST	B.A. OR B.S. (ALMOST ANY FIELD)	B.A./B.S.- IND. MGMT.
DIST. CNTR. CYCLE COUNTER	HIGH SCHOOL DIPLOMA	HIGH SCHOOL DIPLOMA
DIST. CNTR. DELIV. DRIVER	HIGH SCHOOL DIPLOMA	HIGH SCHOOL DIPLOMA
DIST. CNTR. DELIVERY DRIVER	HIGH SCHOOL DIPLOMA	HIGH SCHOOL DIPLOMA

The manpower used in customer and product support.

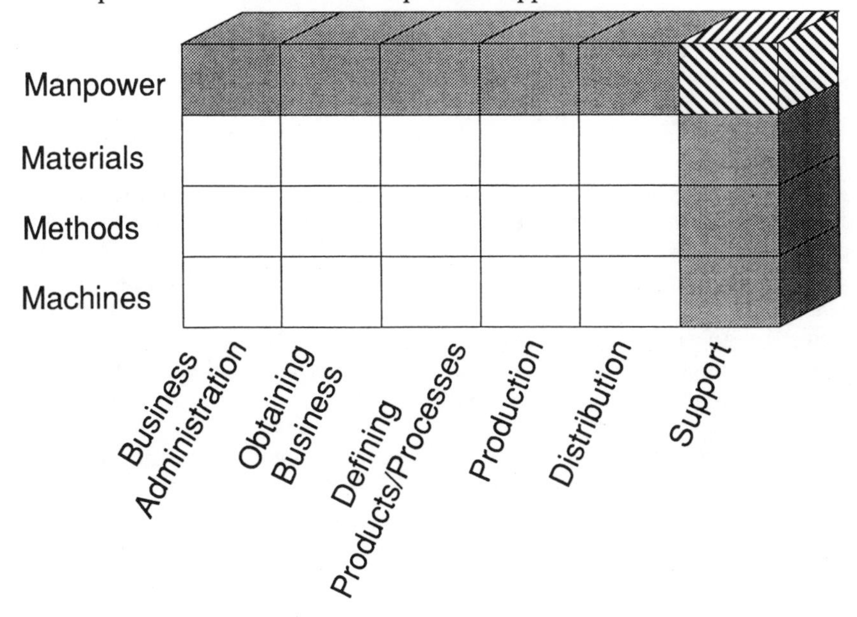

through on-line data bases. This will require computer literacy and a special talent for effectively weeding through data bases to extract relevant information. (If this seems like a minor aspect of employee skills, don't be misled; it isn't. It will result in either cost avoidance or incurred costs in the millions of dollars in connect time and indirect labor hours, and untold losses if the wrong information is used for forecasting.) Also, superior analytical skills will be required to sift through and appropriately apply the data that *is* relevant. This too is an extremely valuable talent, one that will find increasing rewards in terms of wages and status as the information revolution transforms the customer and product support function over the next decade. Skills in computer-based modeling, simulation techniques, and statistical forecasting software will become essential over the next two decades.

Field service repair team training and management will become at the same time easier and more complex. Those manufactured products that are expensive and complicated enough to require field service or repair center service are likely to become more electronically oriented, since the "knowledge content" of these products is increasing. Computers in automobiles, telecommunications devices, home appliances, and other previously low-tech product lines are becoming more akin to higher-end electronic devices in terms of diagnostic and repair protocols. In addition, these products will continue to become more modular in construction over the next decade, making diagnosis and replacement a less time-consuming and labor-consuming process. This implies a command of fairly high tech diagnostic equipment-operating procedures, as well as an ability to comprehend and apply instructions from technical repair publications.

Wanted: Expert System Expertise

A decade or so from now, enough artificial intelligence will be built into our products to allow many of them to diagnose themselves. (Early efforts in this area are already under way. For example, some copy machines currently in use automatically telephone the dealer when they have a problem. Many of them walk users through diagnostic routines when they have a fault.)

These developments will result in a shift in the focus of support professionals away from field repair after the fault has occurred. Technicians will become proficient in anticipating potential problems and designing expert systems that diagnose product faults and lead repair personnel through the restoration process. At that point, the skill set of support professionals will have to include expert systems design expertise. This expertise will prove invaluable not only in the design of diagnostic and repair equipment, but will also be used to make important contributions to the design process, identifying requirements and specifications for vital self-diagnostic circuitry that can be built into the product.

As intelligence becomes a bigger component of manufactured products, the production of technical publications that communicate operating and maintenance instructions to end users will change as well. Already, many products are accompanied by videotapes, audiotapes, and other media rather than traditional printed documentation.

Beyond the evolution to different media, generation of the media will be done differently. Initial changes will require the skills to effectively use desktop publishing software. Later, full-blown multimedia production skills will be required as more and more products utilize audiovisual technical instructions. Finally, the skill requirements will include expert system development skills, as just discussed.

As the product designs become more modular, there should be an improved ability to isolate and replace small modules of the product very quickly rather than going through the extensive disassembly, repair/replacement, and reassembly process typical of most products today. This will be driven primarily by the need to improve repair turnaround times and ensure customer support satisfaction. It will also be enabled to a significant extent by new materials technologies that reduce component weight-to-strength ratios.

Exhibits 11-12 and 11-13 depict likely scenarios for, respectively, the pay and training requirements associated with job titles typically comprising the "support" area in American manufacturing firms over the next ten to twenty years.

Summary: Building a World-Class Workforce

In summary, then, workers in American manufacturing operations will become fewer in number, more technically oriented, better paid, and less interested in work as a way to define their place in the world.

There will be more people available for manufacturing jobs than there will be jobs available—by a wide margin. Those jobs that remain will be better paid, but they will pay higher wages because they require higher levels of education, training, and skill. The (relatively few) people qualified for and employed in those positions will be looking for more interesting, fulfilling work and fewer hours per

Exhibit 11-12. Educational requirements for product and customer support jobs in typical mainstream American manufacturing companies.

PRODUCT AND CUSTOMER SUPPORT
JOBS IN TYPICAL MAINSTREAM AMERICAN MANUFACTURING FIRMS

TYPICAL JOB TITLE	EDUCATION TYPICALLY REQUIRED IN 1990	EDUCATION TYPICALLY REQUIRED IN 2000
V.P. CUSTOMER/PRODUCT SUPPORT	B.A./B.S.-BUSINESS OR ENGINEER.	MASTERS LEVEL-BUS. OR ENGINEER.
CUSTOMER SUPPORT ADMIN.	B.A. OR B.S. (ALMOST ANY FIELD)	B.A./B.S.-BUS. OR ENGINEER.
CUSTOMER SUPPORT REPRESENT.	B.A. OR B.S. (ALMOST ANY FIELD)	B.A./B.S.-BUS. OR ENGINEER.
PRODUCT SUPPORT TECHNICIAN	HIGH SCHOOL W/VOC. TRAINING	ASSOC. LEVEL-MECH/ELEC/CHEM ENG.
FIELD SERVICE REPRESENTATIVE	B.A. OR B.S. (ALMOST ANY FIELD)	B.A./B.S.-BUS., MECH/ELEC/CHEM ENG.
FIELD SERVICE TECHNICIAN	HIGH SCHOOL W/VOC. TRAINING	ASSOC. LEVEL-MECH/ELEC ENG.
TECH WRITER	B.A./B.S.- (MECH/ELEC/CHEM ENG)	B.A./B.S.- (MECH/ELEC/CHEM ENG)
TECH PUB ILLUSTRATOR	HIGH SCHOOL W/VOC. TRAINING	HIGH SCHOOL W/VOC. TRAINING
TECH TRAINING ANALYST	B.A. OR B.S. (ALMOST ANY FIELD)	B.A./B.S.- (MECH/ELEC/CHEM ENG)
TRAINER	B.A. OR B.S. (ALMOST ANY FIELD)	B.A./B.S.-BUS., MECH/ELEC/CHEM ENG.

Exhibit 11-13. Range of wages for product and customer support jobs in typical mainstream American manufacturing companies.

PRODUCT AND CUSTOMER SUPPORT
JOBS IN TYPICAL MAINSTREAM AMERICAN MANUFACTURING FIRMS

TYPICAL JOB TITLE	1990 TYPICAL RANGE OF WAGES ($000 PER YEAR)		2000 TYPICAL RANGE OF WAGES ($000 PER YEAR)	
	LOW	HIGH	LOW	HIGH
V.P. CUSTOMER/PRODUCT SUPPORT	100	150	163	244
CUSTOMER SUPPORT ADMINISTRATOR	40	70	65	114
CUSTOMER SUPPORT REPRESENTATIVE	30	55	51	90
PRODUCT SUPPORT TECHNICIAN	25	45	40	65
FIELD SERVICE REPRESENTATIVE	30	60	48	96
FIELD SERVICE TECHNICIAN	25	45	40	65
TECH WRITER	20	45	30	65
TECH PUB ILLUSTRATOR	20	40	30	60
TECH TRAINING ANALYST	25	60	40	90
TRAINER	25	55	45	88

week. The manufacturing-related work of the future will provide this in spades. Advances in information technology, telecommunications, and materials sciences will provide fulfilling, meaningful work—for someone. There is, of course, the question of how much of that work will be retained here in the United States.

Specific changes resulting from these developments will be smaller, flatter management and support organizations. But the most dramatic reductions will be in direct labor levels. Low-skill jobs that pay a "livable" wage will disappear.

Manufacturing companies will face serious challenges in the development and retention of world-class work forces over the next two decades. Not only will fundamental education levels need to be improved, but advanced skills must be developed in the fields of automation, expert systems development, and other high-tech areas.

The size of the manufacturing organization will shrink dramatically as a result of offloading direct labor, streamlining processes, loss of direct labor as a percentage of the production process through automation, and a general flattening of the organizational structure. Overall, manufacturing organization size is likely to change as shown in Exhibit 11-14.

One other point: The exhibits that depict pay levels and training and education requirements extrapolate likely changes for only those job classifications that exist within the mainstream today. Some of these jobs will all but disappear, and many new ones will be added. Readers should use this information to project what is likely to occur in their own industry and company.

It should be clear that "all things" will *not* "be equal" in terms of organizational

Exhibit 11-14. Anticipated changes in the relative size of the manufacturing organization.

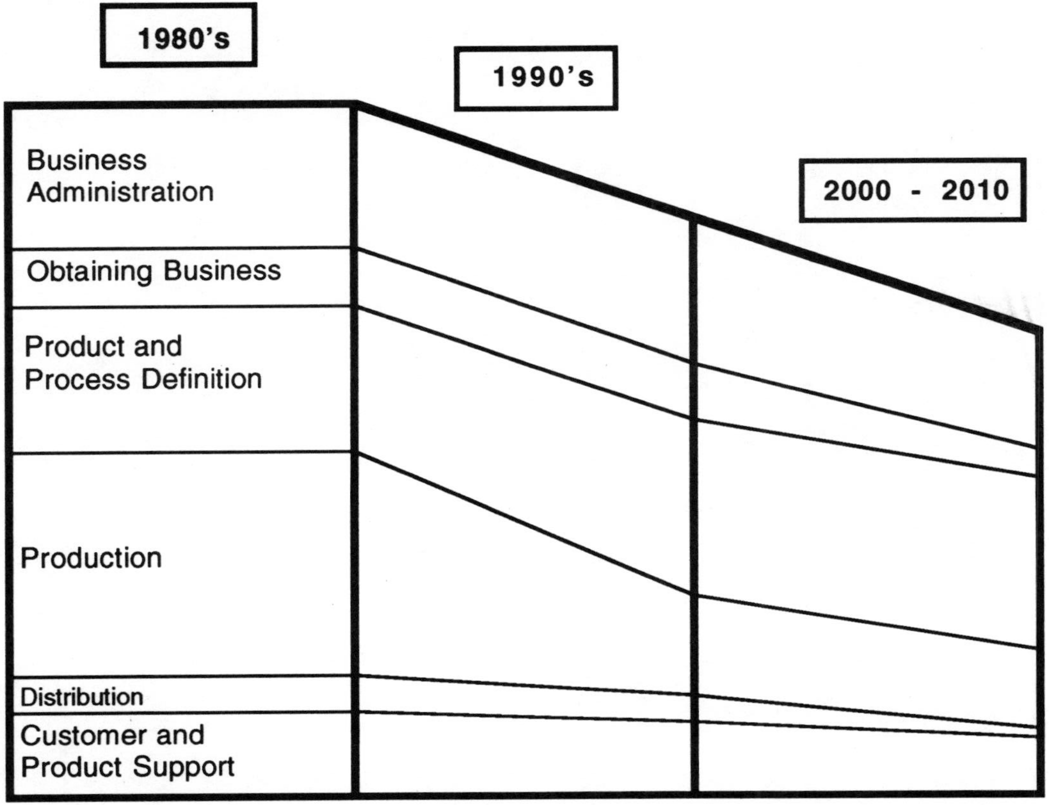

structures or job responsibilities. Since organizational "form" should "follow function," and since the process "functions" will change, we can expect a corresponding metamorphosis in organizational structures and job descriptions. With both the information and the model presented in this book, the reader should have a real advantage in anticipating these developments.

Chapter 12

Dominant Methods: Forging Strong Links in the Global Chain

At their highest level, we can expect the methods used by American manufacturing organizations (see Exhibit 12-1) to evolve in important ways over the next twenty years, including the following:

- Globally linked marketing, design, production, distribution, and support.
- Development and deployment of information tools that filter data, turning it into useful information.
- Diligent exploration and adaptation of the "best practices" of both competitors and noncompetitors.
- Broad, companywide organization structures oriented toward concurrent approaches, particularly concurrent engineering—especially temporary teams formed around new product developments and more permanent teams built around company processes.
- A major shift of all company strategy and activity toward rapid product development and flexible, responsive manufacturing.

Beyond these broad trends, there are myriad specific elements that will appear in future manufacturing methods, and those elements are discussed in this chapter.

The Methods Used in Business Administration

In the area of business administration, we can expect significant changes in the methods used to perform strategic planning, accounting and cost management, human resources management, and management of information processing and telecommunications activity over the next decade.

How Fads Failed Us

Strategic planning methods evolved through a series of fads and theories over the last couple of decades, including management by objectives (MBO), Theory X/ Theory Y, matrix management, systems approach, econometric modeling, decentralization, appliances/partnerships, "intrapreneurship," Theory Z, quality circles,

Exhibit 12-1. Dominant U.S. manufacturing methods.

U.S. Manufacturing Methods

Business Administration	Obtaining Business	Define Products and Processes	Production	Distribution	Support
• Change management organizations • Strategic planning: – Longer horizons – More intense market orientation – Process focus • Accounting: – More timely – Multidisciplined – Activity-based, process-based • Human resources: – Flatter organization structures – More aggressive people – Lower level decision making – More organizations/teams within organizations – More information will be shared with employees – Performance measurement changes from behavior to results, from individuals to teams – Shift from pay-for-position toward pay-for-knowledge and pay-for-performance • Information technology: – Systems will be increasingly reactive or proactive – Proactive IT management will bring: New business opportunities New technology support Company-proprietary technologies – Increasingly outsourced • Factory site selection: – More proximity to markets – More second tier city locations – More, smaller factories	• Market surveys via telecommunications equipment and on-line databases • Expert systems-based market data analysis • Building market analysis findings into strategic plans	• Concurrent engineering (application of expert systems heuristics – integration of SPC data) • New process technology (intelligent processing equipment micro-fabrication – nontraditional processes), GT, cellular manufacturing, FMS	• Production management methods: – Increased percentage of procured parts – Combined MRP and concurrent engineering efforts – Combined production planning, buying, and scheduling functions • Quality assurance: – Movement from reaction to prevention – Broader definition of "quality," broader application of quality tools – More software/less hardware orientation – Widespread application of QFD • Factor floor supervision: – Virtually eliminated • Production methods development: – Increased percentage of "non-traditional" processing – New processes involving composites, ceramics, optic fibers	• Participation of distribution people in concurrent engineering • Pressure to eliminate "middle men": • Move satellite facilities near end markets • Expert systems applications • More outsourcing of distribution work • Automation applications • Heavy reliance on communication standards/proposals • Winners over next 10 years – motor freight and railroads	• Increased emphasis on customer training • Participation in concurrent engineering activity • Heavy reliance on integrated databases and telecommunications systems • Emergence of "infrastructure management"

The methods used in business administration.

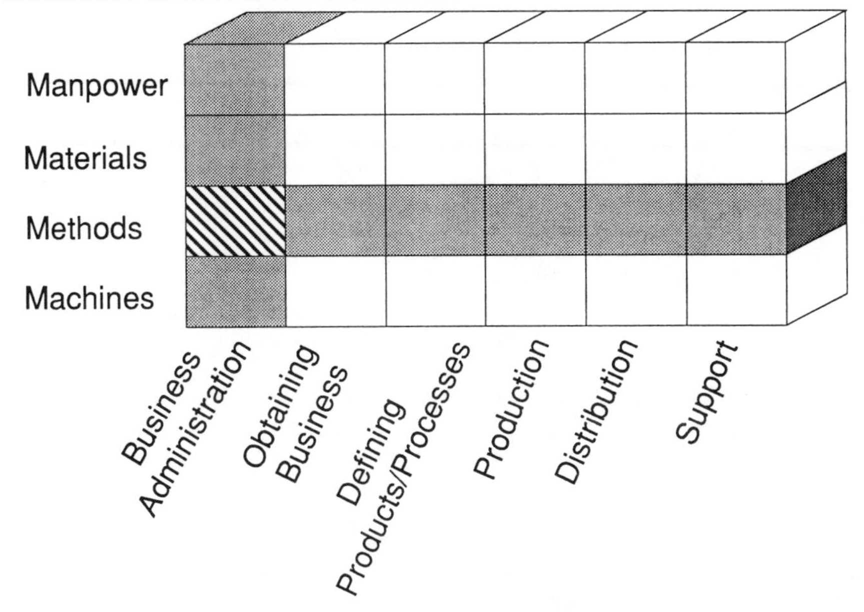

and others. For the most part, the incorporation of all of these new approaches hasn't really served to make us more successful planners. As a matter of fact, some real failures have been racked up over the last twenty years. For example, most new products (accounting for more than half of the money spent on new product development) fail to ever make it to market. Almost half a million businesses filed for bankruptcy in 1988, with more than 300 companies failing in the computer industry alone. As discussed in Chapter 5, the most recent dominant technique in the strategic planning field has been an analytical approach known as SWOT (strength, weakness, opportunities, and threats). It too has been fairly unsuccessful.

In fact, the imitative nature of strategic planning processes in the United States has proved to be a real Achilles' heel. The same approach (however popular and faddish it may be) is unlikely to be best for all companies at all times. Other problems that have served to undermine the efforts of recent strategic planning efforts include these:

- Unrealistic, intangible, or unmeasurable goals
- Lack of executive commitment and accountability to achieving strategic planning goals; lack of "buy-in"
- Repetition, predictability, and inability to break from convention
- Refusal to recognize a failed approach and try something different
- A tendency to rely on size and other "obvious" advantages to carry the day
- Dependency on single product lines, financial sources, and other resources to the point of vulnerability

The fact is, even in the area of strategic planning, most U.S. manufacturers have a great deal to learn from—you guessed it—the Japanese. In a *Fortune* article (April 23, 1990) entitled "Is Long Range Planning Worth It?" Anne Fisher grudgingly admits:

> The unimpressive history of long-range planning in the U.S. would be less vexing if Japanese companies were floundering too. Alas, it seems many of them are succeeding.

She goes on to recount an example of the long-range planning of Matsushita, whose founder developed a 250-year plan in 1932. The plan stated that Matsushita would be manufacturing and selling small appliances in the United States by now, and was right on the mark. Later in the article, Fisher states:

> That the founder was able to imagine the company's far-flung economic might, at a time when Japan was a barely industrialized nation with a tiny GNP, qualifies him for wild-eyed visionary status, right up there with Tennyson.

A few companies have already embraced what is likely to be a very successful approach for dealing with the turbulent years ahead. That approach involves the development and deployment of an overall "change management" process. The purpose of the change management process is to continuously identify, evaluate, and incorporate or defend against changes. The process of change management should resemble the flow depicted in Exhibit 12-2.

Strategic Planning for the Twenty-First Century

Through the next decade, successful strategic planning methods will increasingly require these attributes:

- *Longer time horizons.* Whereas American manufacturing companies traditionally do a five-year plan, our foreign competitors plan in increments of ten, twenty, even fifty years!
- *More depth of understanding.* To be competitive with firms doing this kind of planning, American business administrators will need to ponder and understand the likely advances resulting from enabler fields like quantum physics, anticipating developments in their products, manufacturing processes, and market requirements.
- *More market orientation.* Not merely focused on market share, although that will continue to be a central and essential element, but carefully examining and anticipating market needs.
- *An intense focus on supplier management.* Integrating suppliers from around the world closely into domestic operations will require an unprecedented understanding of other cultures and extremely careful planning. Integrated strategic planning efforts with major suppliers will become an accepted means of developing strategic business alliances.

Exhibit 12-2. The ongoing change management process.

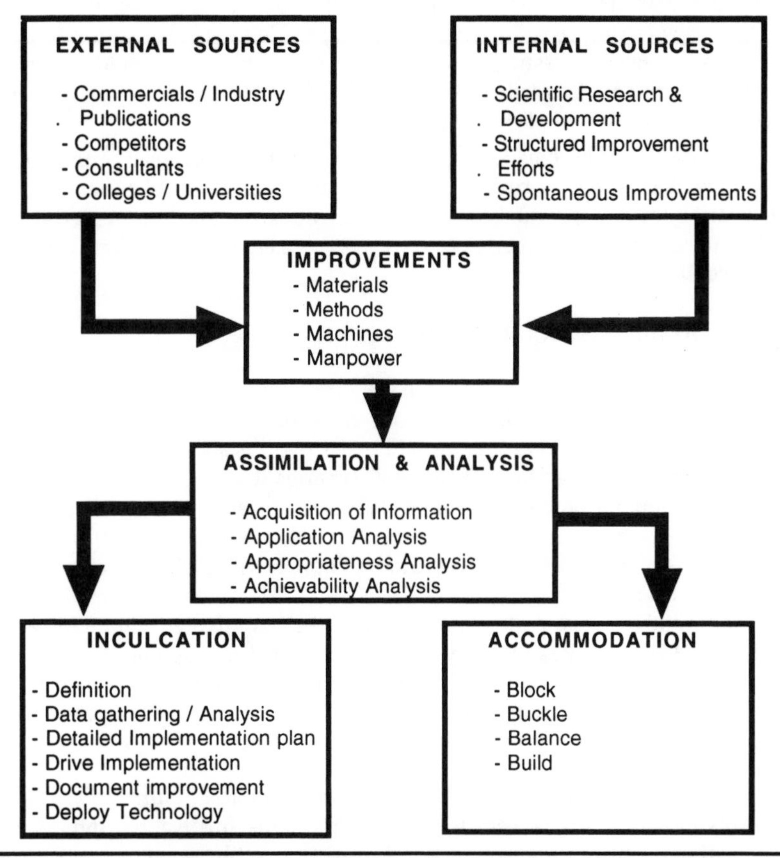

Key to Terminology in ACCOMMODATION Section:

- **Block:** The adoption of competing technologies or strategic moves which prevent the progress of a new technology.
- **Buckle:** A decision not to incorporate the new technology even though we recognize that this will result in the loss of market share. Typically, this decision results from balancing the development and implementation costs agains the potential market segment loss.
- **Balance:** The development of technology which will successfully compete with the one being evaluated.
- **Build:** The development of technology which is complementary to the new one without adopting the technology itself. This is often done when the new technology is proprietary, and the property of a competitor.

• *Accelerated expectations of technological advancement.* Almost no one really understood the rapid acceleration of technological advancement that occurred over the last two decades, but it is now much clearer. The best planners of the next decade will not only account for this kind of change, but depend on it. They will design product, process, and service capabilities around anticipated technology.

• *"Delta team" approaches.* Mirroring the "rapid deployment" groups of our own armed services, successful manufacturing companies will build strategies around multidisciplinary "tiger teams" to rapidly engulf and satisfy newly identified market needs and internal challenges.

• *A (belated) shift from product to process orientation.* American manufacturing companies are beginning to realize that survival depends on the flexibility to produce a variety of new products cost-effectively in extremely compressed times. This will require the development of world-class, "generic" manufacturing and support processes.

When "Creative Accounting" Isn't Enough

Accounting and cost management have (arguably) experienced fewer fad-related phases than strategic planning, but have received much more intense scrutiny in terms of daily operations over the last two decades. Indeed, "creative accounting" tactics are a significant source of earnings in American business, and manufacturing companies are no exception. However, these "earnings" aren't real; they result from changes like moving from "first-in, first-out" (FIFO) inventory valuation to "last-in, first-out" (LIFO) methods. These tactics look great on financial reports, but add no real value.

To be competitive, most American manufacturing companies will be forced to make some dramatic changes in the accounting and cost management areas. The first series of changes for most companies will be designed to improve the effectiveness of existing traditional methods. Many manufacturing companies have traditional accounting and cost management methods that:

• *Do a poor job of estimating costs for new products and new contracts for existing products.* This situation exists because current methods involve redundant and conflicting estimates from different organizational "silos," have no formal process for identifying and incorporating "best cost"/"best value" suppliers in proposals and cost estimation, do not consider technology development projections and other anticipated improvements, and generally rely on outdated historical cost data.

• *Are almost completely reactionary.* Current systems typically don't anticipate changes in either internal operations or external regulatory developments. When these changes do occur, the cycle time to perform impact analyses and actually incorporate the required changes is often dangerously slow.

• *Are tedious and cumbersome even in the normal operating scenario.* For example, fledgling managers in many mainstream American manufacturing companies are frequently shocked to find that they are months into a fiscal year before they really know what their operating budget for that year is going to be.

• *Are neither commonly understood nor user friendly.* It is rare to be able to get consistent (not to mention accurate) responses from several different people (even when they are all executive-level management people) in a given manufacturing company to questions like "What is our current cost of money?" "What is our inventory carrying cost?" and "If we eliminate $100 in excess inventory, how much money does our company actually save?" In fact, it's difficult in many of these firms to get consistent answers from managers to questions like "What percentage of our cost of goods sold (COGS) is direct labor?"

• *Are only beginning to look at internal cost management and product cost estimating as a process that should encompass the entire life cycle of potential and existing products.* Some of this problem has been a result of lack of understanding on the part of managers, but there is also a real need for consumers to understand it better. Managers at major automotive firms faced that problem as they began to get serious about electric vehicles a few years ago. New ways of measuring cost had to be formulated that included previously unconsidered elements in order to establish how a higher up-front investment for an electric car could actually save the car buyer money over several years of ownership (for example, no oil to change, no fluid systems to maintain, and no mufflers to replace).

• *Rely heavily on allocation, particularly as it pertains to assignment of overhead costs.* Direct labor hours are commonly used as an overhead allocation base. This is often misleading, because direct labor usually represents only a small portion of total costs. There is often no logical relationship, and almost never a directly causal one.

Seven Trends in Accounting and Cost Management

As we move through the next decade, improvements will be made in accounting and cost management methods to remedy the problems just discussed, and others will be made to deal with developments such as increased automation. Among the changes likely to occur are these:

• Improved timeliness of reporting actual costs. We will move to virtually "real time" cost accumulation and reporting.

• Improved responsiveness to "what if" questions related to cost impact based on customer requests, technology improvements, and so on through simulation, expert systems application, and other technological advances. For example, "How much would it cost the company to accelerate the delivery of this order so that the customer is satisfied?" (This will be a real competitive advantage in the cycle-to-market area.)

• Multidisciplined cost-estimating "team" approaches. These teams will more quickly assess the interactive cost curves (e.g., price versus volume versus tooling cost) associated with products and services to arrive at a recommended course of action. They will also be much more effective in implementation once strategies and tactics have been determined and it is time to execute them.

• More active involvement of suppliers in integrated processes of cost collection and reporting and cost management, especially in the area of new product development, as strategic alliances and partnerships become an increasingly common theme among American manufacturers.

• Increased emphasis on the make/buy decision. It will be analyzed much more closely and eventually become one of the most meaningful decision-making processes undertaken by management. Accounting professionals will attempt to find some way to give more weight to "core competencies" in this process, at least in the near term (through the year 2000).

• Significant changes in the way we view and analyze investments in plants, equipment, training, and people. Initial focus on return on net assets (RONA) models will lead executives away from capital intensity. Then, as we become more aware of the long-term danger associated with neglecting manufacturing process technology in favor of product technology, the balance will shift. As we saw in our manpower discussion in Chapter 11, most primary business functions will need to become more capital-intensive over the next two decades to remain competitive. Substantial effort is likely to be applied by the financial groups toward methods improvements in managing these costs for profitability over the life cycles of product lines, and over the anticipated life cycles of the process technologies themselves.

• Many of our mainstream manufacturers will continue to pursue activity-based costing in an attempt to derive more realistic general and cost accounting numbers. There is likely to be some divergence here between those who are going for realism and those who will use the allocation method to proceed in a predetermined direction. Those striving for realism will ultimately develop costs and general resource requirements from some device like a "bill of activity." Companies taking this avenue will find it to be a years-long and difficult process. Other companies will favor the approach taken by some foreign competitors, which is simply constructing and manipulating the allocation process to move the company in the direction they choose. An excellent article by Toshiro Hiromoto ("Another Hidden Edge—Japanese Management Accounting") appeared in the July–August 1988 *Harvard Business Review*. Hiromoto states:

> The perspective offered by Hitachi managers seems to be shared by their counterparts at many other companies. It is more important, they argue, to have an overhead allocation system (and other aspects of management accounting) that motivates employees to work in harmony with the company's long-term goals than to pinpoint production costs. Japanese managers want their accounting systems to help create a competitive future, not quantify the performance of their organizations at this moment.

In any event, future cost accounting systems are likely to try to account for non-value-added costs separately from value-added costs, and overhead costs will increasingly be allocated to specific products or services rather than direct labor.

Managing Human Resources in the Flatter Organization

Human resources management promises to be an incredibly challenging field over the next two decades. The technological skills required of the workforce will be greater, the organizational structures themselves will dramatically change, the ways in which manufacturing people will be measured and rewarded will evolve, and the very character of management will be altered.

The workplace itself will change a great deal over the next two decades. Companies are already in the midst of a movement toward flatter, leaner organizational structures. This trend is likely to continue through at least the end of the 1990s. Corporate and staff jobs will be all but eliminated, lower levels of management will disappear, and many jobs currently performed by analysts and coordinators will be replaced with expert systems and other kinds of artificial intelligence software. There will be very few "lifetime" employees, with most workers having ten or more jobs in five or more different companies during their careers.

Because of this and the increasingly competitive environment, the people in the organizations and the organizations themselves are likely to become more aggressive. Decision making will be shifted to lower and lower levels in the organizations. There will be fewer opportunities for advancement, because there will simply be fewer management slots to advance to.

There will be more and more small businesses, and small business "units" within the larger businesses, which will enhance individual autonomy. Employees will be most valued for their creativity, their flexibility and breadth of knowledge, their ability to recognize new business opportunities (particularly related to market "niches") and their communication skills.

More information will be shared with employees than ever before, as they are drawn into a less hierarchical and more participative role. They will be exposed to the following:

- Information about national, local, and market-specific business trends
- All kinds of company performance data
- Incentive program progress reports
- Updates and new developments in the product and service offerings of their company and their competitors
- Research and development efforts, at least in broad terms
- Changes in company strategy and policy
- Updates in benefits program offerings

The reasons this information will be shared are many. Sharing the company's information tends to reduce the level of nonproductive, "rumor mill" activity. It corrects misconceptions, provides constructive feedback on performance, and enhances the "team" atmosphere. This helps break down the barriers and animosity between management and labor ranks. It increases worker concern and involvement, and it is often just plain educational.

Bearing all of this in mind, the way human resources management will be performed over the next couple of decades will change in several ways. Performance appraisals will change, the way performance is measured will change, the means associated with motivating employees will change, the way we pay people will change, and the nature of the behavior we reward will change.

Performance Appraisal

Appraisals will encompass broader goals than ever before, including goals for individual learning, relearning, and skill development. They will often be done at the team level rather than the individual level, and be less formal. They will be conducted increasingly by peers rather than management, and they will be less directly connected to pay.

Performance Measurement

Performance measurement will shift from evaluation of activity or behavior toward quantifying results and specific accomplishments. It will usually focus on teams or "natural work groups" rather than individuals. There will be five to ten specific goals and objectives involved, rather than the current norm of one or two. The goals themselves will be developed by the team rather than management, within the guidelines of overall company strategy. The focus of the goals will evolve from control and efficiency into effectiveness and providing feedback to reinforce good performance or resolve problems. Typical measures will include the categories of quality, financial performance, timeliness, and productivity. There will be lots of feedback. It will be highly visible (lots of charts and graphs, prominently displayed), immediate, and it will emphasize trends in keeping with continuous improvement.

Employee motivation will change as well. The lines between management and the workers will blur, as lower-level managers' and supervisors' jobs give way to team leaders and support groups. The successful motivators will find ways to make work emotionally and mentally appealing to their workers, causing them to feel like a part of a larger purpose or goal. Workers will need to be made to feel like "heroes," in the vein of recent approaches like the ones used by Apple Computer, Lincoln Electric, and Nucor Steel. People will be given a "mission" to "buy into," and be expected to commit to it. They will be "empowered" through employee involvement, and be held accountable for their own team's performance. They will also be motivated by peer pressure, as the team becomes the medium of measurement and reward.

Employee Pay Procedures

The way we pay our employees will increasingly move from pay-for-position to pay-for-knowledge and pay-for-performance programs. In addition, an increasing percentage of our pay (some studies indicate up to 40 percent) will vary from month to month. According to one study described in *Workplace 2000* (Dutton, 1991) by Joseph H. Boyett and Henry P. Conn, the number of firms using small-group incentives and pay-for-knowledge systems will double between now and 2000, with "gain-sharing" programs nearly doubling in the same period. Most new, nontraditional systems will include lower or negligible base salary increases with increasing levels of variable compensation, increasing occurrences of merit-based lump-sum payments, and rewards based on overall company performance.

Gain sharing is a group-based system where employees earn bonuses by finding opportunities to reduce expenditures in areas such as labor, capital, and materials. This kind of approach is projected to grow by 76 percent in terms of its use in the manufacturing sector.

Pay-for-knowledge is a system based on increased levels of compensation proportional to the number of different jobs that an employee can perform competently. It is often at odds with traditional job descriptions, job restrictions, and work rules, and is therefore not generally popular in labor union environments. These systems are projected to grow more than 60 percent in the manufacturing sector.

Employee Rewards and Incentives

The nature of the behavior that will be most lavishly rewarded in manufacturing organizations in the future is leadership-related behavior. More specifically, there will be far less demand for or reward for management, and far more for leadership. Our current management-oriented activity is centered on making existing operations more efficient by reducing cost and compliance with schedules, budgets, and procedures. It is analytical, and relies on tightly controlling the behavior of subordinates to accomplish discrete tasks. Future leadership-oriented activities will be centered more on establishing and communicating a common vision of what the company will be at some point in the future, and inspiring and empowering teams of people to achieve that vision. The skills required of our future leaders in the manufacturing organization include "visioning" skills (more commonly referred to as "farsightedness"), communication skills (both verbal and written), motivational skills, and trust-building and team-building skills.

Increasingly, human resources will be recruiting with a focus on drive, ambition, and other personal leadership characteristics as much as education and training in a specific field. There will be increasing emphasis on rotating current and potential leaders through a variety of jobs in different areas of the company, including international assignments in conjunction with the globalization of both markets and suppliers. There is likely to be an increased emphasis on mentoring and individually tailored training programs. Sabbaticals and leaves for continued education and enrichment are more likely in the future work environment as well. All of these developments will challenge human resources management skills as never before.

Turning Information Systems Into Profit Centers

Management of information processing and telecommunications in the future manufacturing organization will take one of two basic forms from company to company. One form will be a reactive form, where information technology is regarded as a cost center, a purely support-oriented activity. The other form is the proactive form, where information and telecommunications technology management are regarded as potential profit centers, conduits for implementing company strategic plans, and the fabric of communications within the company as well as with the outside world. All this may seem a bit philosophical on the surface. It isn't. The differences are very real, and will strongly affect all companies' financial strength over the next two decades, whether they are manufacturers or not.

In reactive environments, information and telecommunications management will be a low-level function of the organization. It will be unstructured and loosely controlled. This will be an easy pattern to fall into in manufacturing organizations over the next twenty years, for three reasons:

1. Applications for information and telecommunications management technology are increasingly diverse. Information management technology is changing the

face of every facet of manufacturing, from business administration to product and customer support. Personal computers are everywhere and are linked to almost everything through local area networks (LANs) and Electronic Data Interchange (EDI).

2. The technology is becoming more widely available. Costs are coming down and capabilities are escalating very rapidly at the same time. Devices like laser printers and optical scanners, which used to be available only at prohibitive costs, are now only hundreds of dollars, and fax machines are showing up in people's homes and even in their automobiles. Laptop computers, cellular telephones—the list is endless.

3. The technology is increasingly user friendly. Spread sheets, word processing software, desktop publishing software, data base management software, and flow-charting and general graphics software are all in widespread use in virtually every company. They are sometimes on stand-alone desktop or laptop computers, sometimes resident on LANs, and sometimes reside in a mainframe with "dumb terminal" access.

In the reactive environment, the management information systems (MIS) manager will be so caught up in reacting to the daily integration and development needs of all the individual user communities that he or she will never gain real control. The most this individual can hope for is to develop a formal process for prioritizing and allocating resources to system development efforts.

In this environment, data integrity and data security are an absolute nightmare. Typically, the MIS manager reports to the vice president of accounting, or perhaps elsewhere lower in the organizational structure. The individual is seldom if ever involved in the strategic planning efforts of the company, and would be ill prepared to do so, because he or she operates at an almost purely tactical level. The extent of strategic planning in this situation is determining which of next year's user requests will fit into next year's budget.

In the proactive, profit-center environment, the picture is quite different. The head of MIS (sometimes called the chief information officer, or CIO), reports to the president or CEO and is part of the executive board. He or she is a vital part of strategic planning activities, and has a strong voice in decisions pertaining to the future allocation and expenditure of capital. This individual is charged with looking some years into the future and developing integrated, overall information management processes and systems that not only support the rest of the manufacturing organization but also enhance the profitability of the company.

Strategic planning in this environment will point to areas where the proper management of information and telecommunications technology can accomplish the following:

- *Bring new business opportunities to the company.* This will be accomplished in a myriad of ways; for example, performing design work for other noncompeting companies at a profit to fill unused design engineering capacity, and transferring that design work directly via EDI to the customer. Or offering the application of state-of-the-art expert systems for everything from market analysis to design producibility to outside, noncompeting businesses at a profit.

• *Assist in the introduction of new technologies.* This will be achieved by making the expert systems, on-line data base information, and computer aided instruction of noncompany experts available inside the company through EDI and on-line information services, and by offering new packaged software evaluation and introduction.

• *Development of proprietary synergistic technologies.* Because of their role in the strategic planning activities of the company and because of the integrated approach to development, insightful MIS leaders will be able to recognize system development opportunities that will satisfy more than one apparently isolated future development need while making all of them better (for example, development of an expert system for product configuration that also provides an improved method of presenting products (selling) to potential customers).

Critical Challenges for MIS Leaders

In both reactive and proactive environments, MIS leaders will face several critical challenges over the next two decades, including the following:

• The question of whether computing services should be purchased from an outside firm, and, if so, which services should be "farmed out" and which should be kept. Some major American manufacturers and service companies have decided to virtually turn over all computer operations to outside companies. Companies providing this service include EDS, IBM, Andersen Consulting, DEC, Computer Sciences, KMPG Peat Marwick, and AT&T. U.S. businesses spent more than $7.2 billion on outsourcing these operations in 1990 alone. According to a *Fortune* article entitled "Why Not Farm Out Your Computing?" (September 23, 1991), half of America's major corporations are evaluating whether to convert to this kind of operation, or have already decided to do so. Among the ones who have are Kodak and Cummins Engine. Various companies have cited savings from this conversion of 90 percent on capital spending, 10 percent to 20 percent on operating costs, and an overall annual savings on their data processing function of around 40 percent.

• Overall systems integration approach. This would appear at first to be less of a problem for the MIS manager in a reactive environment. After all, this individual has no long-term plan to achieve total systems integration. However, what this really translates to is an environment where decisions about interface and integration methods and tools will have to be made over and over again, building endless bridges between diverse systems. In the proactive environment, there will be several high-level, protocol-related decisions to be made about the connectivity between largely already existing (MRP II) systems, CAD/CAM/CAE systems, accounting systems, and so forth. Additional interface and integration work will need to be done in small applications written by users themselves, although this will become more of an oversight function as software becomes increasingly "friendly" and users become more proficient.

• Data security and intercompany electronic communications protocol development. The primary defense in the 1990s against computer crime of all kinds, including general data security problems, is heavily restricted access. Unfortunately, this is seldom an effective approach. It makes new product or system development cumbersome, which decimates both productivity and morale. Especially in the environment of trust and empowered teams that we will be trying to foster over the next two decades, this will be a tough approach to hang on to. Once it has occurred and been discovered, the theft or malicious alteration of company data is also an extremely difficult thing to prove and trace to a specific culprit. Beyond these difficulties, at least thus far, companies have been reluctant to report the loss of important technical data or the compromise of financial data to the police, and thereby to shareholders. It is likely that a significant market will develop around the private investigation and resolution of computer crime by the year 2000.

In terms of system structure, in this decade the basic approach to develop secure commercial networks will include the following tactics:

- Networks connected to master files will be physically secured, with leased lines or on-site lines that belong to the user.
- Access to the master file network will be controlled by password and biometric identification (e.g., fingerprint, voiceprint).
- Data encryption standard (DES) devices (composed primarily of a single computer chip for encoding or decoding) will be installed at each end of these communication lines.
- Master file processors will contain access control software that verifies the authorization of specific terminals and specific requests.
- Heuristic audit trail-monitoring software will track usage patterns and learn to distinguish regular from irregular activity.
- Master files will be kept in encrypted form, with DES links between master files and all file management subsystems and file access subsystems.
- Gateway processors will separate the master file network from all file access networks, where another set of encryption and heuristic audit trail processor devices will be installed.
- Access through uncontrolled links will be isolated from controlled networks and closely monitored for unauthorized inquiries, viruses, and other anomalies.

Some progress has been made to date on electronic communications protocols, such as ANSI (American National Standards Institute) X.12 and AIAG (American Industry Action Group) standards. With the introduction and widespread use of fiber optics, ISDN (Information Systems Data Network) is likely to bring increasing complexity to this issue. Standards will need to be developed, incorporated, and upheld for voice networks, data networks, digital multiplex systems, open systems interconnections, and any telesystems used.

According to Norman Weizer in his book *The Arthur D. Little Forecast on Information Technology and Productivity* (Wiley, 1991):

> Two sets of standards will be followed by all vendors in the mid-1990s that supply complete IIS systems: the IBM SAA, SNA, DIA, DCA de facto series, and the ISO Open System Interconnection (OSI) model. All seven levels of OSI standards will be implemented by the mid-1990s. A

significant number of other minor standards will be enforced by groups such as the apparel and auto industries and the government.

Computer-aided acquisition and logistics support (CALS) will be the most widely implemented standard for compound document exchange (CDE). In 1990, the Department of Defense required all contractors to comply with CALS as specified in the 1840A Standard, which establishes digital interface between computers exchanging information necessary to support weapons systems.

Another CDE standard, the Office Document Architecture—Office Document Interchange Format (ODA-ODIF), will be in early stages of acceptance in U.S. markets. With backing from the ISO, ODA will gain support in Europe first. Vendors such as DEC endorsed the ODA concept early because it allowed for the incorporation of video and voice.

Perhaps the greatest challenge of all for our future MIS leaders will be overcoming the attitudes created by disillusionment and disappointment in their efforts over the last 20 years. In spite of the claims of most software vendors, the promises of MRP II and CIM (computer-integrated manufacturing) proponents to date have, in many cases, gone unfulfilled. They have proved too costly, required too long to implement before tangible improvements could be realized at a "bottom line" level, and simply been too complex to be well understood by senior management.

Commuting on the Electronic Highway

Still another area, related to all of these areas we've discussed under business administration, is likely to have a significant impact on us over the next decades—telecommuting. Telecommuting is performing work for the company at home, branch offices, or other remote locations, "commuting" electronically (rather than physically) to the office. There are many advantages and opportunities associated with telecommuting, but many risks as well.

Telecommuting grew out of the convergence of computer technology advances, value changes of today's employees, and changes in the attitudes of corporate management itself. It offers advantages in recruiting, hiring, and retaining talented staff. It also offers increased productivity opportunities, along with reduced asset requirements (that's improved RONA for accounting types). However, it requires different kinds of management and may involve the "political" hurdles of reorganization, as well as some significant effort in identifying and "packaging" the work so that it can be done effectively from a remote location. There are also challenges related to regulation, data security liability, and compensation and benefits. A very good analysis of all of these factors can be found in the book *Telecommuting* by Gil Gordon and Marcia Kelly (Prentice-Hall, 1986). In terms of growth, Gordon and Kelly estimate that the potential telecommuting population is about one-fourth of the American workforce.

How MIS Methods Affect Factory Site Location

A final concern related to business administration methods over the next couple of decades is factory site selection. This matter is tied to the explosive information technology revolution, because of improved information availability throughout the company, irrespective of individual factory locations.

Since management information systems will provide virtually instantaneous reports of remote facility performance, the strongest need for centralization has been eradicated. Recent emphasis on customer satisfaction as a measure of overall quality and striking interest in managing to the shortest possible cycle-to-market have combined to create an environment where the tendency is to locate multiple (smaller) factories closer to the end product consumer. Chapter 2 of *Toward a New Era in U.S. Manufacturing* by the National Research Council states:

> In fact, there is some evidence that, due to responsiveness, flexibility, and quality concerns, future trends in factory locations, particularly for component manufacturers, will be toward a proliferation of smaller factories closer to final markets. . . . For some industries, the concept of the microfactory will become important: small factories, highly automated and with a specialized, narrow product focus, would be built near major markets for quick response to changing demand.

Likewise, a "Plastics Survey" published in *Plastics News* (March 19, 1990) showed that 60 percent of the companies responding indicated that "proximity to end market" was very important.

General trends have been movement toward second-tier cities, out of the rust belt of the Midwest toward the southern reaches of the United States, and, of course, off shore. However, some offshore companies have recently moved manufacturing operations back to the United States. The reasons include rising comparative overseas wage rates, productivity gains here, quicker market response, faster product development and cycle-to-market, poor operating results in overseas efforts, and just plain old patriotism.

Companies that have returned various manufacturing operations from overseas to domestic sites include American Tourister, Chicago Pneumatic Tool, GE Fanuc Automation, Tandy, Texas Instruments, and Xerox.

Overall, then, it looks as though the smart money is on small factories located near their end markets. (A good example is Motorola's pager manufacturing, which involves manufacturing facilities in Singapore for the Asian market and in Florida for the U.S. market.)

The Methods Used in Obtaining Business

Methods used in obtaining business through the 1990s will be broadly affected by changes in customer values and better identification and incorporation of customer needs into the producer's operations, products, and services. Another substantial influence will be advances in marketing support technology.

Consumer Values

The values of the American consumer have changed dramatically over the last two decades. The consumer of the 1990s is much more demanding than ever

The methods used in obtaining business.

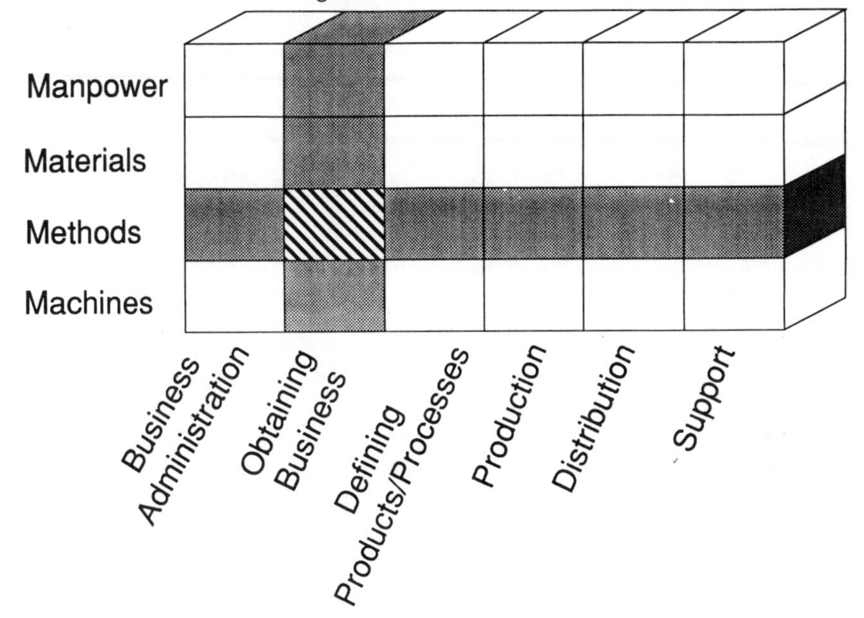

before. People are asking for customized services and products suited to their individual needs and tastes. They want uncompromising quality at "world-class" standards, and those standards are increasingly being set by our foreign competitors. Value, as measured by the ratio between quality and price, would have to be regarded as the most important overall element. When consumers are surveyed, quality and price almost always come out at the top of the list of factors they look for in making the purchase decision.

Customer Needs—Catchwords for the Next Century

Satisfying customer needs means, of course, that marketing and sales efforts will need to shift from an emphasis on "Look what a great product this is" to "How can we develop products and associated services that satisfy your special needs?" It will mean tying the marketing, sales, R&D, design, and distribution functions of the manufacturing organization all into the process of defining and delivering customer satisfaction. There will be a continuing movement away from identifying new applications for existing products toward tailoring products and developing a broader range of specialized ones. Specifically, marketing leaders will be required to do the following:

- *Identify, document, and prioritize customer needs.* Market surveys and studies, the long-standing tools of the profession, will become far easier to do through the use of on-line data bases and other high-tech tools. However, for a thorough assessment of customer needs there is no real substitute for the face-to-face interview. Technology will help here in terms of gathering and correlating the market survey data and communicating the data back to the manufacturing facility in nearly real time. Developing and deploying these methods and taking the best possible advantage of the new technologies will be a major factor in the success of marketing and sales professionals throughout this decade.

• *Analyze market and competitor data to identify market niches, emerging trends in customer needs, and emerging strengths and weaknesses in competitive position.* Again, newly developed systems technologies and others still in development will be useful. Particularly helpful will be expert systems that track customer data, prospective customer data, sales history, and correspondence and develop market-niche-specific sales approaches to satisfy those niches. Many existing software packages, including "ACT!" by Contact Software International, the "Brock Activity Manager Series" by Brock Control Systems, and "Contact Ease" by West Ware, have been on the market for some time. These packages do most of the customer data management, sales history tracking, and report-generating functions related to marketing and sales. However, most of the software related to identifying sales trends, strengths and weaknesses, and recommended sales approaches is still proprietary, and most of it has a long way to go to reach its potential. Such proprietary software will appear throughout the middle 1990s and probably through the end of the decade. Marketing leaders will face the question of whether to develop these expert systems based on what they perceive to be the unique aspects of their own businesses, or try to go with a "vanilla" package offered by an outside vendor. Some will elect to do both, building "bridges" between a sound vanilla package and internally developed expert systems to perform the more exotic analytical functions.

• *Incorporate the findings into the company's strategic plan.* It will be vital to the success of the long-range marketing effort that intelligence gained during the upstream marketing and sales efforts is effectively built into decision making related to product designs, manufacturing capabilities, and distribution and support structures. The interaction of the marketing and strategic planning processes will evolve into a much more responsive process, with monthly and even weekly iterations in most companies by the year 2000. Sales forecasting will rely less on the current trending methods (e.g., exponential smoothing, least squares, linear regression) and be more forward-looking. As the market environment grows more volatile, historical data loses more and more of its value.

The next two decades will witness market volatility in most of the markets for American manufactured goods. Here is some pertinent advice given by Diane Sanchez, president of sales consulting group Miller Heiman, in the May 1992 issue of *Success*:

> Stake out specific opportunities for new business rather than setting goals for new business and then trying to figure out a way to meet them.

For many American manufacturers, this will be a real departure from business as usual in the 1980s.

Beyond these broad activities, there will be other, more specific changes in the methods involved in obtaining business for manufacturers in the years ahead. For example, there will be a great deal of emphasis on strategic alliances between noncompeting companies in order to perform less resource-intense marketing and to make "packages" of products and related services available to customers for leveraged sales, greater margins, and increased customer satisfaction. These alliances will also be used to open previously closed or heavily restricted markets, a factor that will increase in importance throughout the next two decades as the market becomes increasingly global.

The ongoing proliferation of special interest organizations for every conceivable profession, industry, and interest will also prove to be an increasingly fertile area for marketing professionals. Already, management consultants and software salespeople are among the most frequent speakers at seminars and regular dinner meetings of professional organizations like the American Production and Inventory Control Society (APICS) and the National Association of Purchasing Management (NAPM). Technology will continue to provide a foundation for the development of even more of these organizations (and resulting marketing opportunities) through the development of user groups for popular software packages. For example, a clever manufacturer of hydraulic tubing might find out what kind of manufacturing software his major customers (potential and current) use, then buy it also. By incorporating this software, even in some token fashion, the manufacturer gains access to the user group, can develop relationships with potential customers, and can even use the new system (which is "common" between his own operation and his customers') as a sales feature, offering to share commonly formatted data electronically with the customer. Thus, he streamlines the ordering process, delivers the tubing in the sizes and quantities called out in the customers' own systems, and gains a perceived edge over competitors who utilize different systems.

Finally, there will be differences at the most tactical level. For example, marketing professionals in the 1990s will continue to utilize technology like the fax machine. But increasingly, they will distribute sales information on new products and product-related services to customers and targeted potential customers through fax "broadcasts." Increasing access to and familiarity with these existing technologies will change our businesses as much as the "new" technologies we adopt over the next several years.

The Methods Used to Define Products/Processes

The methods used to define products and processes during the next two decades will likely center on two concepts: concurrent engineering and new process technology implementation.

Concurrent engineering, like MRP, CIM, and just-in-time (JIT), is often given a unique name at each different company. (This must be done to foster ownership and avoid what is often referred to as the "not invented here" syndrome.)

Concurrent engineering has been defined as "the integration of product and process design." It usually involves a physically colocated, interdisciplinary team including specialists from product design and manufacturing engineering. As we move through the 1990s, these teams will expand to more commonly include

The methods used in defining products/processes.

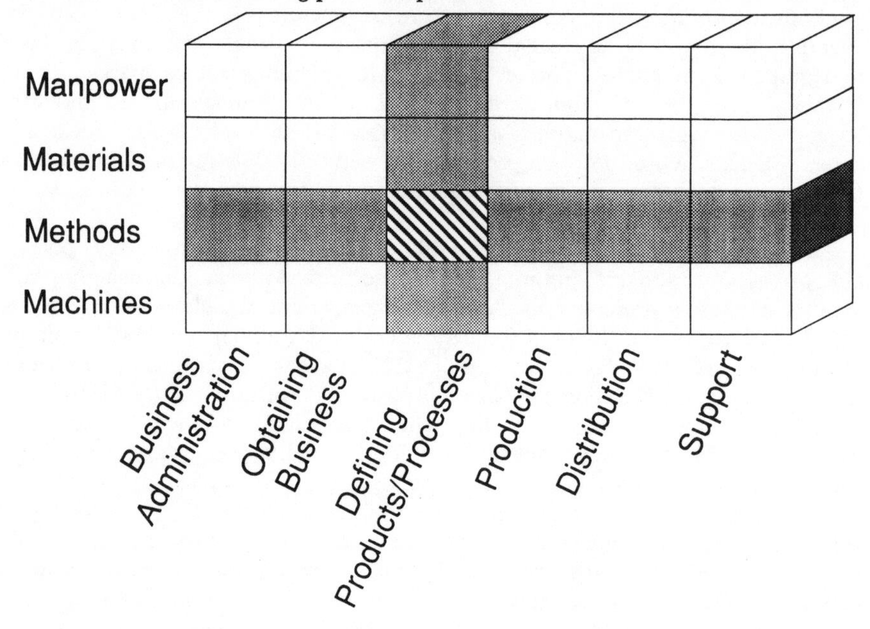

representatives from manufacturing, marketing, and distribution as well. However, far more than physical colocation is required.

A Winning Combination: Concurrent Engineering

The companies that will be real winners in the area of concurrent engineering in the 1990s will be the ones that recognize and act on this axiom: Process design is as important as product design, and both are most effective when they are performed simultaneously. The companies that fully grasp this principle will be those that recognize that all supporting processes are involved here, not just production processes. Manufacturing lot-sizing, machine and work center routing, tooling development, packaging development, handling and distribution method identification, and myriad other processes will eventually be incorporated into this activity. However, the simultaneous development of manufacturing processes and products will be as much as most companies will achieve in the 1990s. The balance of this growth will most likely occur through the first decade of the next century.

There are two reasons this will take so long: First of all, there are some technical bridges that, while already well along, must be completed in most companies. These "bridges" are really the final development and eventual integration of software packages that design manufacturing processes. To be effective, this software must recognize the individual process capabilities unique to the manufacturing environment for which they are designing. This implies relational data bases containing the details pertaining to all of the elements affecting the capability (and therefore the expected quality) of each process (e.g., operator skill level, machine repeatability, tool repeatability, and material characteristics).

The next several years will see the development and frequent deployment of expert systems that maintain this data and recommend production process designs on the basis of the data retained in these relational data bases.

The next step will likely be integration of statistical process control data fed back automatically and on a real-time basis from the production machines, allowing the relational data bases to update themselves.

Eventually, heuristics will be applied to allow the production process planning system to "learn" from the experience of the combinations of operators, tools, machines, and material how to fine-tune design criteria and predict "yield" quantity from any given combination. This will be an important advance for predictive costing of proposed designs and all other cost/pricing work. We will probably achieve substantial deployment of this kind of system by the close of this decade.

Around the turn of the century, then, as we develop reliable systems in this area, the scope of concurrent engineering activities will dramatically widen, encompassing the supporting processes mentioned earlier in this section. The reason several other disciplines will be added almost simultaneously is that the model for this type of system will have been developed and proved for production processes. It will be a development pattern readily emulated in the other areas. For example, the critical elements used for packaging might include packaging material types, packaging equipment requirements, and packaging manpower and skill requirements. Again, the repeatability (and therefore output quality) of each element will be measured, monitored, and tracked in relational data bases. And so it will go, emulating the success of the production process design system in developing a packaging process design system.

The second reason that the development and widespread deployment of full-blown concurrent engineering processes will require nearly fifteen years is that a "paradigm shift" will be required on the part of practically the entire manufacturing organization. As my friend and colleague John Kramer is fond of pointing out, "You can get there technically long before you can get there mentally." The problem has been defined as "paradigm shift" (*Discovering the Future* by Joel Barker), "culture shock," "changing the mind-set," and in some other less-pleasant terms. Anyone who has tried to implement real change in old-school manufacturing organizations will readily recognize what I mean.

Our concurrent engineering leaders in the twenty-first century will manage a process that iterates between virtually all the major processes of the company as they develop and evolve with the addition and updating of products and services (see Exhibits 12-3, 12-4, 12-5, and 12-6). The concurrent engineering process will be almost "chemical," bringing together the diverse "valences" of design, financial, and other systems and controlling the application of group technology, material requirements planning, statistical process control, and so forth, like a chemist would apply heat or pressure in the laboratory.

Inroads in Adaptable Process Technology

A 1988 research study conducted jointly by the Society of Manufacturing Engineers (SME) and A. T. Kearney, Inc., was commissioned to explore the future

(*Text continues on page 178.*)

Exhibit 12-3. The concurrent engineering process.

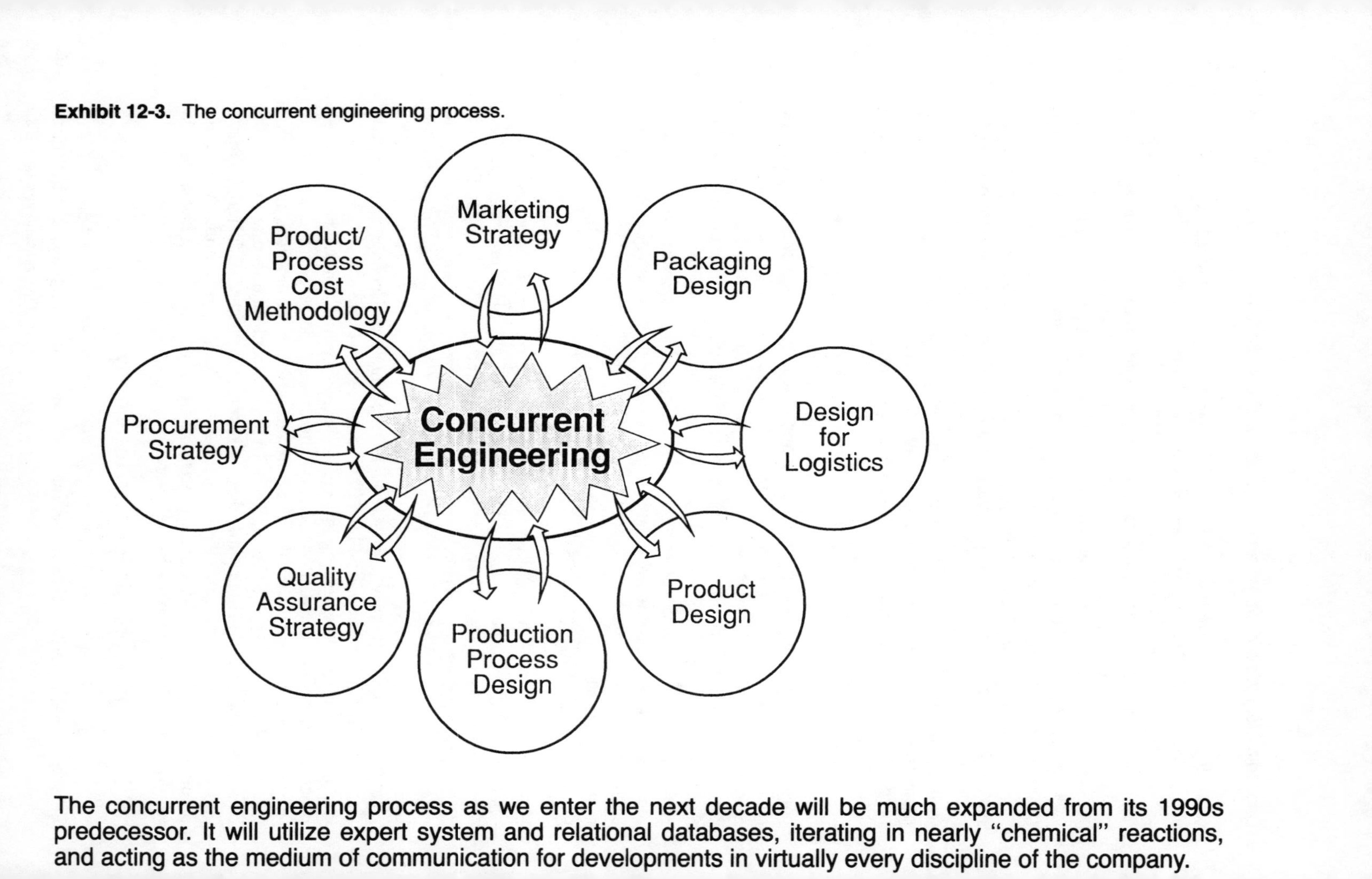

The concurrent engineering process as we enter the next decade will be much expanded from its 1990s predecessor. It will utilize expert system and relational databases, iterating in nearly "chemical" reactions, and acting as the medium of communication for developments in virtually every discipline of the company.

Exhibit 12-4. Comparison of traditional product design and concurrent engineering.

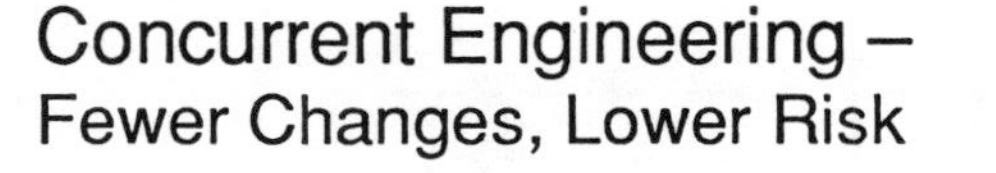

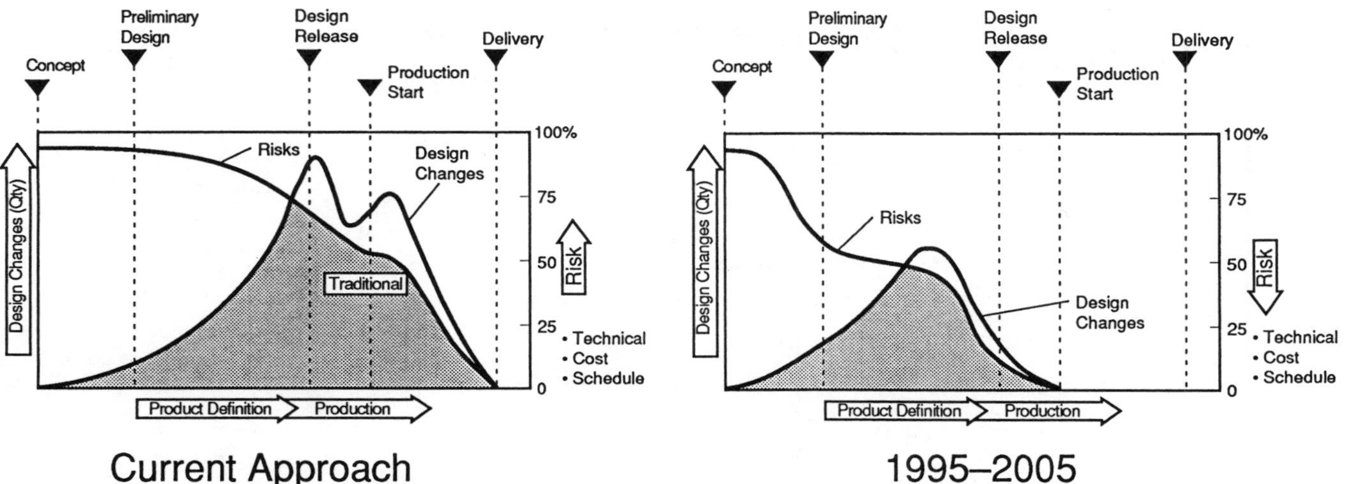

As we move through the 1990s, concurrent engineering is enabling us to drive risk levels to earlier and earlier stages of the product development life cycle.

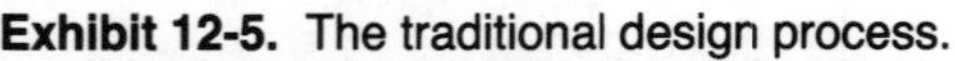

Exhibit 12-5. The traditional design process.

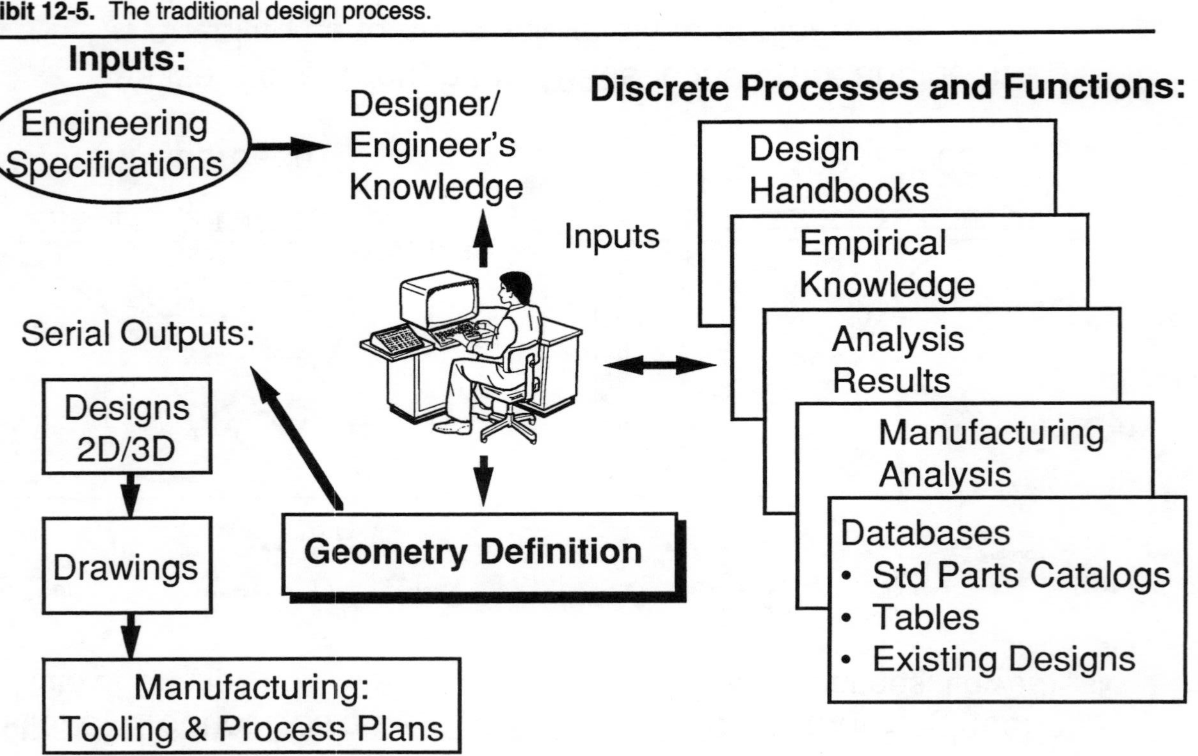

Exhibit 12-6. Knowledge-based engineering process.

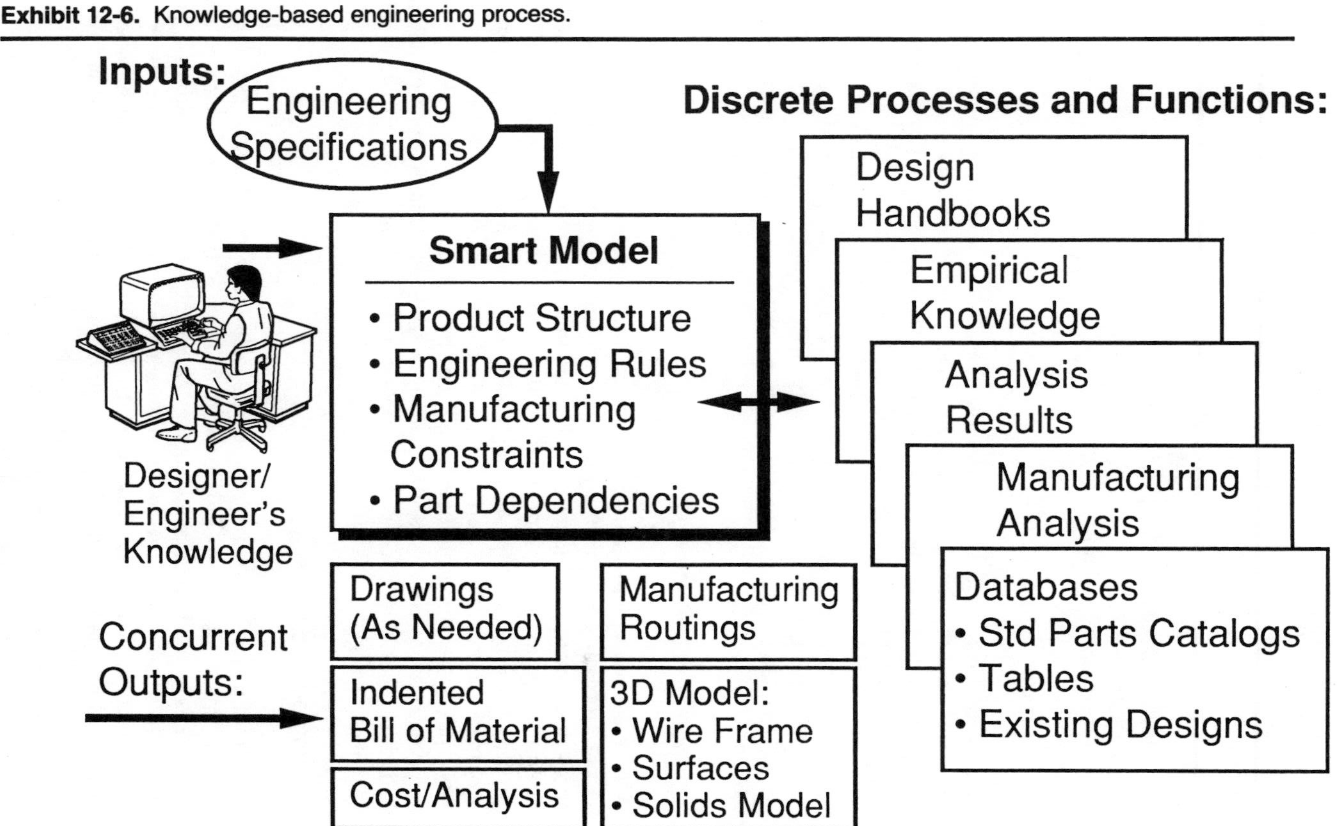

role of the manufacturing engineer. The study findings were published in a report entitled "Profile 21: Issues and Implications." In it, the authors recount the comments of roundtable participants who describe, among other things, their views about products of the future. Key elements of future products, according to this report, including the following:

- More frequent product changes
- More "simple to operate" products
- More feedback features
- Fewer components and new materials
- Improved quality
- Greater recycling

This study further reflects that the role of the manufacturing engineer is likely to shift away from the design of a manufacturing process that mass-produces the same items toward designing a process that can easily be adapted to manufacture multiple products on the same equipment at competitive cost.

In terms of new process technology implementation, we will see significant changes in a number of areas. Clearly, the sequential convergence of technologies like artificial intelligence, robotics, and advanced materials (as described in our model in Chapter 8) will manifest itself in important ways throughout the next two decades.

Intelligent Processing Equipment

One major development in process technology application is in the area of intelligent processing equipment. This equipment utilizes sophisticated sensors to monitor critical material properties as the materials are altered during production operations. Later, as artificial intelligence is applied to these pieces of equipment, the processes will be designed and operated in a manner that allows the management of material properties. As we move into the first decade of the next century, the first widespread applications of intelligent processing equipment are likely to occur in mainstream American manufacturing facilities. This equipment will include industrial robots fitted with advanced sensor equipment and driven by intelligent control systems.

Microfabrication

A second major development in process technology implementation will be in the area of microfabrication. This process will grow out of the convergence of new materials technologies, biotechnology, artificial intelligence, simulation, and robotics. The process will involve manipulating and fabricating materials at the microscopic level. The first likely industrial application of this technology is production of integrated circuit (IC) chips utilizing the microscopic application of films and surface preparation materials. (There will be metallurgical applications such as minimum-friction bearings as well, but they will likely follow the IC applications by some months, if not years.)

This field will finally yield public entry into an entire science dubbed "nanotechnology," which will allow us to create "designer" materials by constructing and altering them at the molecular and, eventually, atomic levels.

The field of nanotechnology was first brought to public attention by K. Eric Drexler in a fascinating book entitled *Engines of Creation* (New York: Anchor Press, Doubleday, 1986). A more recent book on this same subject by Drexler, Peterson, and Pergamit entitled *Unbounding the Future* (New York: Morrow, 1991). In this work, the difference between future molecular manufacturing and today's production machining processes is described as similar to "the difference between watchmaking and bulldozing."

Nanotechnology probably will not begin to become widely available as a medium for manufacturing until around 2010. However, microfabrication activities are likely to emerge beginning around the turn of the century.

Techniques on the Threshold

Yet a third process technology advance involves a group of miscellaneous processing techniques that will reach threshold levels of capability during this period. These include material formulation techniques and genetic engineering methods related to "designer" materials, multidomain smart sensor technology and application, solid-state laser technology, optics technology, photonics technology, and directed energy technology.

In addition, some near-term developments have already been deployed in limited applications and will become more widespread over the next ten years. Among them are these:

- Nontraditional advanced machining processes, including the following:
  - —Mechanical processes, such as ultrasonic machining, abrasive jet machining, and water jet cutting
  - —Electrical processes, such as electrochemical machining and electrochemical grinding
  - —Thermal processes, such as electron beam machining, laser beam machining, electrical discharge machining, and plasma arc machining
  - —Chemical processes, such as chemical milling and chemical blanking
- Nontraditional advanced bonding processes, including the following:
  - —Advanced adhesives, such as high-temperature adhesives, high-strength adhesives, expanded elastic range adhesives, and rapid-cure adhesives
  - —Advanced welding processes, such as electron beam welding and laser welding
  - —Molecular surface bonding processes, such as diffusion bonding
- Automated traditional and (eventually) nontraditional processes, applying automation techniques and robotics to traditional fabrication and assembly processes; and later, expanding these applications to include the nontraditional processes described previously

Exciting Times Ahead

Finally, those developing and defining production processes over the next ten years will certainly see the expansion of several leading-edge production process approaches, such as group technology (GT), cellular manufacturing, flexible manufacturing systems (FMS), and computer-integrated manufacturing (CIM).

The first three, GT, cellular manufacturing, and FMS, will be especially useful in high-volume operations that produce families of parts having similar characteristics. CIM will be a prerequisite for manufacturers as automation becomes a predominant aspect of their production operations. However, the uniqueness of individual automated systems continues to make CIM a complex and challenging goal, and will likely remain a substantial deterrent to CIM through the end of the decade.

Because fabrication and assembly technology will evolve in these ways and others over the next two decades, the methods we use to define production processes while doing concurrent engineering must evolve as well. Through the balance of the 1990s, we will have to learn to assess the capabilities of new production process technologies and incorporate them more rapidly than ever before, reflecting these capabilities in new computer-based expert systems that can apply heuristics and identify optimum processes for new products. Additional software developments will need to occur toward the end of the 1990s that allow us to anticipate process development requirements and identify potential process variation levels that could generate nonconforming product specifications (defects).

As we enter the first decade of the next century, we will need to apply the systems developed for these purposes in the 1990s to microfabrication and later to microassembly. Simulation technology will be very useful in the design of these systems, and we will test the existing limits of our knowledge of artificial intelligence as well.

All of this points to very exciting times ahead in terms of the methods used by mainstream American manufacturers for product and process definition.

The Methods Used in Production

The methods used in production were summarized in Chapter 5 in two broad categories: production management methods and actual conversion methods. That same approach is used here as well to discuss changes likely to occur in this area over the next several decades.

Production management methods in mainstream manufacturing environments include procurement, production planning and control, quality assurance, and shop floor supervision.

The New Power of Procurement

Procurement is likely to take on increasing significance as American manufacturers enter the twenty-first century. In a 1988 study of American manufacturing entitled *Countdown to the Future: The Manufacturing Engineer in the 21st Century*, A. T. Kearney and the Society of Manufacturing Engineers published the results of a survey of CEOs from major American manufacturing firms. Here are some survey responses:

- More than half of the respondents currently bought more than 20 percent of their components from outside sources.

The methods used in production.

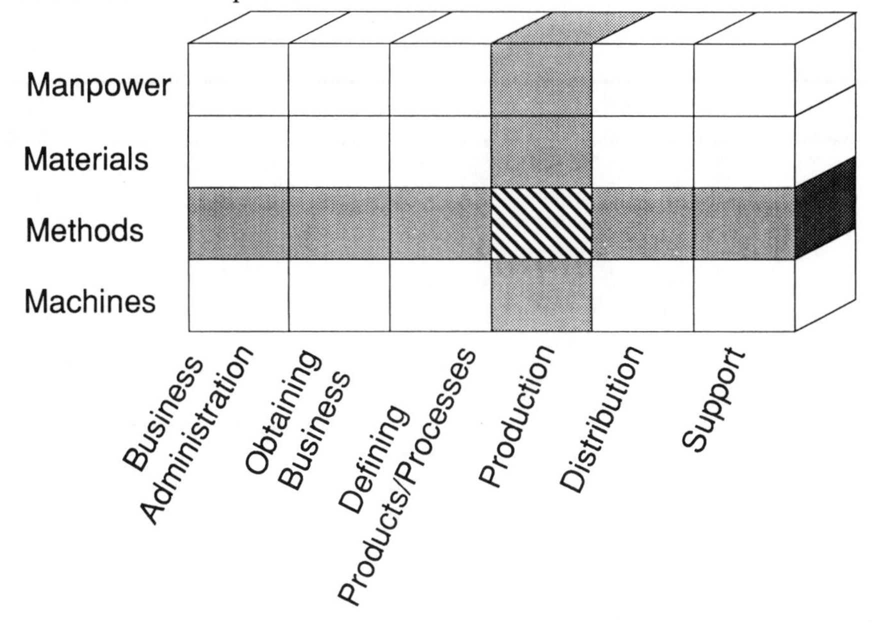

- All the manufacturers responding that currently bought more than 20 percent of their components from external sources planned to buy an even larger percentage of them by the year 2000.
- Virtually all those who currently bought more than 20 percent of their components from foreign manufacturers expected that percentage to approximately double by the year 2000.

Given this trend, then, we can expect that procurement will be an increasingly vital aspect of production management. Procurement is likely to change in a number of ways over the next decade, including the way it is organized, the skills needed by buyers, and the way it is approached as a result of the increase in purchased components.

Organization

Procurement professionals are currently organized in a manner that reflects the orientation of the business. For example, manufacturing organizations that use large quantities of similar parts will generally have a cadre of procurement professionals who concentrate on a specific commodity such as forgings, sheet steel, resin, or paint. Other manufacturers will procure primarily major components such as engines, fuselages, and transmissions. In these cases, there are often procurement specialists who focus on these specific items.

Over the next eight to ten years, the procurement organization in mainstream

American manufacturing organizations will probably evolve away from these traditional structures. In the commodity-rich environments, procurement organizations will likely move toward a centralized negotiating team for the commodities used to maximize flexibility, responsiveness, and quality while minimizing price. This team will be composed of negotiating experts and have temporary members who specialize in the individual commodities, as they move from commodity to commodity. Once the order has been negotiated, the follow-up and day-by-day execution of the contract will be turned over to the computer. Reorder point (min-max) systems, consignment inventories, and actual part consumption are likely to provide the triggers to suppliers for replenishment in these cases.

In the major subcontract item environment, it is likely that the procurement specialist will remain associated with the particular commodities, but again will relinquish actual expediting and de-expediting functions. However, in this environment, those activities will be more closely monitored to ensure against advance deliveries (because they represent a much more significant portion of product cost) and are likely to be handled by material liaisons on the factory floor whose job it is to maintain constant communications with the supplier, accelerating and decelerating the actual material deliveries as required. Both parties will also closely monitor the subcontracted item in terms of delivery, responsiveness, and product quality.

Skill Set

As we move through the balance of the 1990s, buyers in many manufacturing organizations will be required to broaden the scope of their responsibilities to encompass planning functions, fulfilling what is increasingly known as the "buyer/planner" function. This will entail all of the job responsibilities between the receipt of Material Requirements Planning (MRP) output from the master schedule through negotiation of the purchase contracts. It will eliminate some jobs, but it will improve the continuity of the materials management process and streamline its throughput time.

In addition, the skill set of the buyer will need to include not only the skills required today but also skills that allow him or her to remotely manage some of the supplier facilities involved in the production of the manufacturer's products.

Approach

In the face of globalizing supplier bases and increasing percentages of purchased components, the procurement professional will evolve from a buyer to a manager of remote manufacturing. Procurement contracts of the future will be structured much differently to minimize total cycle-to-market and maximize flexibility of configuration. For example, many purchase agreements will be written to procure a percentage of the supplier's capacity and authorize procurement of long-lead raw materials at the supplier's facility. The agreement will specify the capacity in terms of machines, tools, operators, and materials to be available at the supplier's, with close-in determination of exact part configurations and incremental delivery quantities. For example, if the supplier makes three different engines that require most of the same material, equipment, and manpower to produce, the contract might stipulate that 1,000 engines be purchased from the supplier during the next calendar year. However, the purchaser might reserve the right to specify the mix of the engines as late as thirty days from expected delivery. This allows the supplier to put long-lead items in work and "lock out" enough capacity each week to produce an annual quantity of 1,000 engines.

Transitions in Production Planning and Control

Another mainstream production management method, planning and control, will, of course, evolve as well. MRP (material requirements planning) systems will still be utilized, but it is likely that the overall scheme of MRP II (manufacturing resource planning) will be folded in to a large extent with the concurrent engineering effort, and that both of these will be enveloped in whatever future manufacturers choose to call their individual CIM (computer integrated manufacturing) programs.

Initial attempts to analyze the impact of new products and processes on existing capacity (tooling, machines, manpower, and facilities) will no doubt eventually be worked into the concurrent engineering effort, rather than being largely ignored until design work is completed, as they so often are today. The convergence of MRP software capabilities and simulation capabilities with the other tools supporting concurrent engineering will provide the foundation required for these changes.

The job of production planners in current MRP environments is to analyze and manipulate the output from the MRP system, converting it into purchase requisitions and manufacturing orders. As we go through the 1990s, these folks will find that their jobs have been combined with those of specialists in either procurement or shop floor scheduling. New job titles will be "buyer/planner" and "planner/scheduler." In both cases, the changes will serve to reduce redundancy, enrich the jobs of the people involved, and improve accountability for the quality of the materials management functions of the organization.

Shop floor production control in the more progressive organizations will all but disappear. The primary functions of shop floor production control people are to prioritize the work in each work center and expedite part shortages. In an environment where part families are all produced in manufacturing cells or flexible manufacturing systems, there is very little prioritizing required. These facilities work basically on a first in, first out (FIFO) basis. This is especially true when the facility is converted to a "pull" system consistent with just-in-time operations. Once the bugs have been worked out, there is simply no additional prioritizing to do. Part shortages in this environment will also be an anomaly. Safety stocks of low-value items will be calculated and maintained by the MRP system, quality levels will improve to the point where most of the nasty little "surprises" we often experience today will be eliminated, and the balance of the problems resulting in part shortages will simply not be tolerated. In a real just-in-time environment, part shortages are deadly. They get a lot of attention, and are generally resolved very quickly. These are lessons that are still being learned by U.S. manufacturers, sometimes at great cost (i.e., the recent strikes at G.M. facilities which shut down Saturn production lines). But they are being learned, and we can expect the constant flurry of production control expediting activity to diminish dramatically as a result.

The Question of Quality

In the next two decades, quality assurance, a third dominant production management method, will be an area of tremendous opportunity for those who offer the proper skills and approach, while affording almost nothing to those who don't. Unfortunately, the majority of our current QA workforce falls into the latter category. The typical approach to quality assurance in the United States has been almost completely reaction-based. It has focused on product inspection and corrective action. The skills required included a knowledge of how to read micrometers, calipers, and, in the more exotic environments, sophisticated defect detection equipment such as X-ray machines and automatic ultrasonic inspection systems.

Even now, QA is coming to mean something much different. The emphasis has shifted in many companies from reaction and correction toward prevention. Statistical process control (SPC), Taguchi methods, and other techniques have helped us refocus our efforts. It is becoming increasingly clear that we can no longer delegate the responsibility for the quality of our products and services to an inspector after the fact. Everyone in the manufacturing organization is responsible for the quality of their work.

Beyond this new focus on prevention, the very definition of quality in many organizations is growing from the traditional "meeting all design specifications" toward "delighting the customer." TQM (total quality management) programs are springing up everywhere, and with the advent of the Baldrige award, quality is increasingly the focus of national attention. Another powerful force in unifying and crystallizing quality guidelines more uniformly between companies, industries, and countries is the set of international quality standards known as ISO9000. Drafted by the International Standards Organization, this set of standards is a promising start toward a dramatic improvement in developing a common language for QA professionals, and a worldwide benchmark against which manufacturing organizations can measure themselves, their suppliers, and their competitors. Such standards are rare in any segment of the manufacturing world today, probably because they are so difficult to construct in a fair and comprehensive manner. ISO9000 is a pretty good start, and not a bad model for other disciplines in terms of how to develop this kind of guideline.

Quality assurance professionals in the future will operate in two realms. The primary realm won't be on the factory floor at all; it will be on the concurrent engineering team. The quality engineer's primary role will be to ensure that the product and processes designed are fail-safe, that they can't produce a defect. This, of course, while minimizing cost. It will be an enormously important job and require skills not often present in the QA organization today. Ensuring that processes are fail-safe will be the key. It will increasingly require an intimate understanding of all production-related processes and the ability to exercise control over those variables in each process that produce critical characteristics. It will involve turning virtually every employee into an expert about his or her own process quality characteristics, and training them in maintaining optimum quality levels throughout those processes. Later in the product cycle, it will evolve into an audit function, systemically monitoring the processes to ensure consistent capability levels to prevent the potential for defects.

In general, quality control will involve more software and less hardware. The mechanics of monitoring product quality and ongoing process quality should be almost completely automated in mainstream manufacturing facilities by around 2005. When any process moves toward a control limit, the monitoring system involved can alert the operator, then the area management before any potential for defect arises. However, the development of these systems will take several years yet, and once they have been developed, the individual processes that are added must be evaluated, documented, and maintained. All of this is computer-based work and will require skills quite different from most of those in our current QA organizations.

The other major realm of QA operation will be customer service. By means of QFD (quality function deployment), a primary responsibility of the quality professional will be to identify, interpret, and convert customer needs into critical product characteristics. Then the quality engineer will map the critical product characteristics to critical process parameters and ensure that those parameters fit within existing or planned process capabilities. (Again, this role fits neatly into the overall concurrent engineering process.) Finally, it will be the responsibility of the quality engineer to design mechanisms into the processes to promote fail-safe operations, minimizing the opportunity for defects to occur.

Supervising the Shop Floor

The fourth mainstream production management method, factory floor supervision, is another job classification generally considered an "endangered species." In fact, the number of supervisors required per unit produced will decline dramatically by the turn of the century, primarily because the amount of direct labor required per unit produced is going to decline dramatically. Automation and the current drive to outsource increasing percentages of our components, especially to foreign suppliers, are largely responsible. In addition, the skills required will change. Future first-line factory management will require facilitating and coaching skills, because the direct labor employees will increasingly be organized into teams mirroring the group technology–based application of manufacturing cells and flexible manufacturing systems. It will not be enough to be the best machine operator in the group to qualify as a supervisor. Future supervisors will need to be leaders, not merely technically or physically skilled workers.

Actual Conversion Methods: Techniques Follow Materials

The second of the two major categories of methods used in production is the actual conversion methods. Developments in these methods over the next two decades will closely follow the developments occurring in materials. They will not be evenly distributed and uniform in their arrival, but rather will cascade in surges from improvements in material processes. These material process developments will spring from advances in chemistry and physics, supported by breakthroughs in the physical handling of materials at the microscopic (and later, the molecular) level.

Most of the really dramatic changes in material processes are most likely to occur around the turn of the century. However, the material composition of many of our products is already changing quite rapidly. With these new materials (primarily composites, ceramics, and optical fibers), changes have come in the kinds of equipment and production processes utilized. Other drivers of material conversion processes have been computer technology and simulation.

Next, we will review some of the interesting developments under way, which should help construct the context for speculating about what will happen in the near future. We can't begin to cover them all here, but a brief overview should suffice for our purposes.

Metalworking Processes

Metalworking processes will likely include the traditional stamping, forming, drawing, and spinning processes currently in use through the end of the 1990s. They will also continue to include traditional forging and extrusion technology, as well as the most common machining processes (e.g., turning, drilling, milling, broaching).

However, we will see an increasing number of nontraditional processes supplanting these traditional methods as we reach the end of this decade. Particularly in the metal-machining area, we will see the increasing emphasis on material conservation and close tolerances create an elevated level of interest in nontraditional processes. These processes are not necessarily new. Many of them date back decades, and in a few cases even centuries. However, it has only become practical to apply them in many cases as the combination of quality requirements and enabling technology has finally converged.

Among these nontraditional machining processes are the following:

- *Ultrasonic machining,* which has successfully been applied to carbides, stainless steels, ceramics, and glass. This process utilizes the high-frequency oscillation of a cutting tool in abrasive slurry, with the tool shaped to correspond with the shape to be produced.
- *Abrasive jet machining,* which has been successfully applied to germanium, silicon, mica, glass, and ceramics. This process utilizes the effect of fine abrasive particles suspended in a high-velocity gas stream directed at the material in work.
- *Abrasive flow machining,* which has been successfully applied primarily to metals for polishing, deburring, and radiusing radial cutting/finishing operations. This process uses a pressurized flow of viscous, abrasive-laden liquid across the material surface.
- *Orbital grinding,* which has been successfully applied to metals and graphite. This process employs an abrasive "master" that abrades its full three-dimensional shape into the workpiece.
- *Water jet cutting,* which has been successfully applied to nonmetallic parts such as kevlar, glass epoxy, graphite, boron, and fiber-reinforced plastics. This process uses a fine, high-pressure, high-velocity jet of water.
- *Electrochemical machining,* which has been successfully applied to hard metals that are too difficult for more traditional machining methods. It applies a reverse plating action to remove metal from a positively electrically charged workpiece through anodic dissolution.

• *Electrochemical grinding,* similar to electrochemical machining except that it applies a combination of electrochemical action and abrasion to remove the metal. It also utilizes electrically conductive workpieces, as well as a grinding wheel composed of an insulative abrasive set in a conductive bonding material.

• *Electrochemical discharge grinding,* which is similar to electrochemical grinding except that the metal removal is principally caused by electrolysis via AC or pulsating DC electrical charges and the process uses a graphite wheel rather than an abrasive one. It is effective on virtually all electrically conductive materials.

• *Electron beam machining,* which is used primarily to cut very small holes and slots in thin-gauge materials. This process utilizes high-power beams of electrons, moving about one-half the speed of light, to bombard and vaporize the material in a focused area of the workpiece.

• *Laser beam machining,* which is used to drill holes in and cut all metals, plastics, paper, rubber, ceramics, composites, and crystalline materials. This process amplifies and focuses an intense beam of light on the workpiece to create high-speed ablation, removing material by vaporizing it and forcing some of it to run off in a liquid state at high velocity.

• *Electrical discharge machining,* which is used to machine electrically conductive materials. This process basically uses sparks to machine materials.

• *Electrical discharge grinding,* which is also used to machine electrically conductive materials, also uses repetitive electrical sparks to achieve the removal. However, it differs from electrical discharge machining in that the sparks are discharged through a gap between a rotating wheel and the workpiece, flushing particles away in dielectric fluid.

• *Plasma arc machining,* which has been applied successfully to brass, bronze, nickel, tungsten, aluminum, mild steel, alloy steel, carbon steel, stainless steel, copper, cast iron, molybdenum, magnesium, and titanium. This process utilizes a constricted arc to melt the material of the workpiece, then blows the molten metal out of the kerf with a high-velocity jet of ionized gas.

• *Chemical milling,* which has been successfully applied to aluminum and aluminum alloys, beryllium, brass, bronze, cobalt alloys, copper, gold, lead, magnesium, molybdenum, and nickel. This process involves the removal of metal from selected surfaces, primarily to lighten the parts, by exposing unmasked portions of the surfaces to etchant solutions.

• *Chemical blanking,* which has been successfully applied to copper and copper alloys, nickel-silver alloys, magnetic nickel-iron alloys, steel, aluminum, magnesium, molybdenum, titanium, and selected plastics. This is the process of exposing selected surface areas on a sheet of material to a chemical etchant solution, dissolving all of the material except the finished part.

• *Chemical engraving,* which has been applied to nearly all metals and has proved most successful and useful on aluminum, brass, copper, and stainless steel. This process utilizes masking materials and chemical etchants to produce finely detailed etched images for products such as name plates.

Composite Production Processes

Production using composite materials has grown substantially over the last decade and will continue to grow throughout the foreseeable future.

Composites are materials composed of a reinforcement material (typically fibers, particles, or whiskers) suspended in a matrix or binder material. They offer several advantages over the metals they replace, including specific tensile strengths of four to six times that of steel and aluminum, modulus (stiffness) levels three and one-half to five times that of steel and aluminum, and higher fatigue endurance limits than both those metals. In addition, they offer the advantages of lower corrosion potential and simpler fastening methods.

There are, of course, obstacles to be overcome, but they are falling every day. The primary problem has been the skill set of the process definition engineers involved. In the foreword of an excellent technical book on composites entitled *Fundamentals of Composites Manufacturing* (SME, 1989) by A. Brent Strong, Stewart Luce, the chairman of the Composites Group of SME, notes:

> Many automotive engineers are duly impressed with the strength and stiffness of composite laminates. But the mastery of new design and production concepts is paramount to the full utilization of composite materials. If increased flexibility in manufacturing is to be realized, today's engineer must wear the hats of designer, chemist, and manufacturing planner, and must develop a keen understanding of the automated factory.

Although the principles of composites production (time, temperature, and pressure) have existed longer than mankind has existed on this planet, once again we find that the proper elements of technology and need didn't finally converge until recent decades.

Now, with the advent of composite components in such industries as automotive, aerospace, computers, construction, artificial limbs, electrical equipment, marine equipment, and space structures, there can be little doubt about the need to understand and improve the production processes associated with composites.

Composites manufacturing methods include the following:

- *Manual lay-up.* In this process, a fabric or mat that comprises the reinforcement material is saturated with liquid resin (which comprises the matrix), and the lay-up is made by building up layer upon layer on a shaped surface (or mold) to achieve the required thickness.

- *Pre-preg material lay-up.* This process allows a generally superior product to be made (as compared to manual wet lay-up methods) with reduced amounts of resin and material-handling labor. It utilizes a reinforcement material preimpregnated with resin (typically available in tape, sheets, and rolls) that is cured slightly to increase resin viscosity. The pre-preg material is cut to the shape of the mold and built up in layers of this pre-preg material to the required thickness.

- *Automated tape lamination.* This process utilizes automated tape-laying equipment to lay up the parts. It can cover flat surfaces and both simple and gentle compound contours. Much of this equipment is still in development, but several pieces are actually in productive work today.

- *Cutting of uncured materials.* This process utilizes a wide variety of equipment, including manual knives with replaceable blades, automated reciprocating knives, ultrasonic vibratory cutters, die cutters, laser cutters, water jet cutters, and broad-goods-cutting machines.

• *Vacuum bagging.* This process applies a vacuum to assist in compressing the plies of material to promote compaction and volatile fluid/gas withdrawal. It involves placing a vacuum bag around the laid-up material on its mold and attaching a vacuum port through the only unsealed opening in the bag. The vacuum is typically applied immediately prior to curing, and often remains in place through the cure process.

• *Autoclave curing and bonding.* Autoclaves are basically pressurized ovens that allow heat and pressure to be applied simultaneously for prolonged periods to the lay-up. The application of these elements causes the chemical reaction that produces molecular "chaining" in the resin and forms a single part from the multiple plies of material.

• *Filament winding.* In this process, a continuous tape composed of resin-impregnated fibers is wrapped around a mandrel by machine to form the part. Often, the resin impregnation occurs as part of the winding process, rather than using pre-preg tape.

• *Pultrusion.* This is a continuous processing method that involves impregnating continuous reinforcement fibers, pulling them through a die, and curing them. (It is ideal for making pipes and similar products having uniform cross-sections.)

• *Matched-die molding.* There are three types of matched-die molding. One is preform molding, which is accomplished as follows:

- A dry mat of reinforcement material is shaped (preformed) and placed onto an open lower mold.
- Resin is added to the mold.
- The upper half of the mold is put into position, and the mold is heated to cure the part.

A second type of matched-die molding is sheet molding of compounds, in which sheets of resin-impregnated chopped reinforcement material (typically glass fibers) are loaded into compression molds, where they are pressed and cured. The third type of matched-die molding is bulk molding of compounds, in which a doughy mixture of chopped reinforcement material (e.g., fiberglass), resin, initiators, and filler are weighed out and placed in molds for pressing and curing.

• *Resin transfer molding.* (Sometimes called resin injection molding.) In this process, a mold is loaded with reinforcement material, the mold is closed, and then resin is injected into it with a vacuum drawing the resin through the reinforcement material.

• *Spray-up methods.* These processes involve spraying a mixture of resin and chopped reinforcement material into a mold, then hand-rolling the mixture in the mold to extract any trapped air bubbles and compress the material onto the mold surface.

• *Cutting, drilling and machining.* The processes used to cut, drill, and machine cured composite parts are more closely related to grinding than cutting because of the abrasive nature of reinforcement materials. With some modifications, traditional metalworking equipment such as mills, lathes, saws, and routers can be used. (Carbide and diamond-tipped tools are recommended, and lubricants may be required during these processes to avoid overheating and resin build-up.) Water-jet cutting is increasingly popular, as are laser cutting and diamond wire cutting.

- *Bonding and joining*. This process is accomplished by two methods. The most popular method of joining composite parts with other composite parts, metal surfaces, and other materials is adhesive bonding. The other, less popular method is mechanical fastener application.

Adhesives are the most popular method because they generally have lower stress concentration, minimize the possibility of delamination, which so often occurs during drilling for mechanical fastening, weigh less, and are less expensive. Epoxy adhesives and rubber-based adhesives are most commonly used. (The most critical aspect of adhesive bonding is surface preparation, which typically involves the application of solvent or vapor degreasing.)

Another method of adhesive bonding in composite parts is welding (or fusion bonding), in which the surfaces to be joined are heated to a melting point, then pressed together. The heating for this process may be produced by conventional means or by induction, friction, or ultrasound.

Mechanical fastening of composite parts involves the application of rivets, bolts, pins and blind fasteners through holes drilled in the parts. Mechanical fastening is most commonly used when peel loading is a critical element, and when there may be a need to remove the part at a later time.

Ceramic Production Processes

Production uses of ceramics are already far more extensive than most people recognize. Because of their ability to resist heat, resist wear, and resist chemical degradation, they have become almost indispensable in many applications. Electrical applications include inductor cores and recording heads. Optical applications include phosphors, lenses, and lasers. Thermal applications include insulation, heat exchangers, and molds. Structural applications include turbine components, seals, and cutting tools.

The actual production of ceramic material is quite simple. The elemental powders of the ceramic material are mixed in water to form a slurry that may be dried in a mold to virtually any shape, then fired in a kiln to achieve hardness. What alters the physical properties of ceramics (e.g., hardness, ductility, etc.) is primarily the chemical elements included in their composition. For example, ceramics with high melting temperatures are typically composed of strongly bonded metals or are "multivalent ionic" ceramics. Low thermal expansion is achieved by using materials such as lithium aluminum silicate or fused silica.

There are, however, some fairly new processes for increasing toughness of ceramics and enhancing some of their other properties. Many of these are still in development and will become widely available only in the late 1990s. The most important area along this line is the development of methods that increase the toughness of ceramic materials. Among the methods in development are second-phase dispersion and fiber reinforcement. Second-phase dispersion involves the addition of small amounts of ductile cobalt and nickel, providing for redistribution of applied loads and thereby minimizing cracking. Fiber reinforcement is the process of reinforcing glass-ceramic and ceramic matrices with high-strength silica-carbon fibers.

Optic Fiber Production Processes

Another major production method is that used to manufacture optic fibers. We are all now irrevocably influenced by optic fibers in our daily lives. They are at the heart of almost every telecommunications system, having proved their superiority to conventional copper wire in virtually every application. They are an integral part of cable television, broadcast television, and remote monitoring and surveillance systems. They are especially well suited to the transmission of digital data such as the data generated by computers. Military applications include communications, aircraft command and control links, satellite earth station data links, and command post communication links. The advantages of fiber optics in these applications include low weight, EMI rejection, and lack of signal radiation. Generally speaking, optical fibers are small, light, and much more efficient than the metal wire they replaced.

Optic fibers are most commonly formed by the "double crucible" method. In this method, one crucible is placed inside another with a conical opening in the bottom of each, something like the working end of the ink cartridge in a ballpoint pen. (The crucibles are generally made of platinum.) The inner crucible is filled with molten core-glass and the other crucible contains molten cladding-glass. The two glasses come together as they are dispensed through the opening at the bottom of the chambers, forming a glass-cladded core. This mixture is "pulled" as it is dispensed, forming a fiber.

Optic fibers may also be produced by pulling them from "preforms" made through processes such as external vapor deposition and axial deposition.

Again, many of the techniques involved in producing and enhancing optic fibers are still in development. It seems reasonable to conclude that the production and application of these materials will remain an increasingly vital aspect of the American manufacturing scene for decades to come.

The Methods Used in Distribution

Distribution methods of the next two decades will continue to encompass three primary areas: internal stocking and controls, logistics, and distribution center management. However, the approaches used in these areas will change.

The most fundamental change will be rooted in the participation of distribution professionals in the concurrent engineering process. As a participant in the concurrent engineering team, the distribution professional will not only make meaningful contributions in the decisions surrounding the (traditional) areas of packaging and material handling, but will also help to determine total distribution strategy, including elements such as factory site locations and vertical integration levels by facility along the value chain of the manufacturing company.

For example, through the late 1980s, distribution channels were organized into two categories: industrial goods and consumer goods. Industrial goods include those distributed in the following ways:

The methods used in distribution.

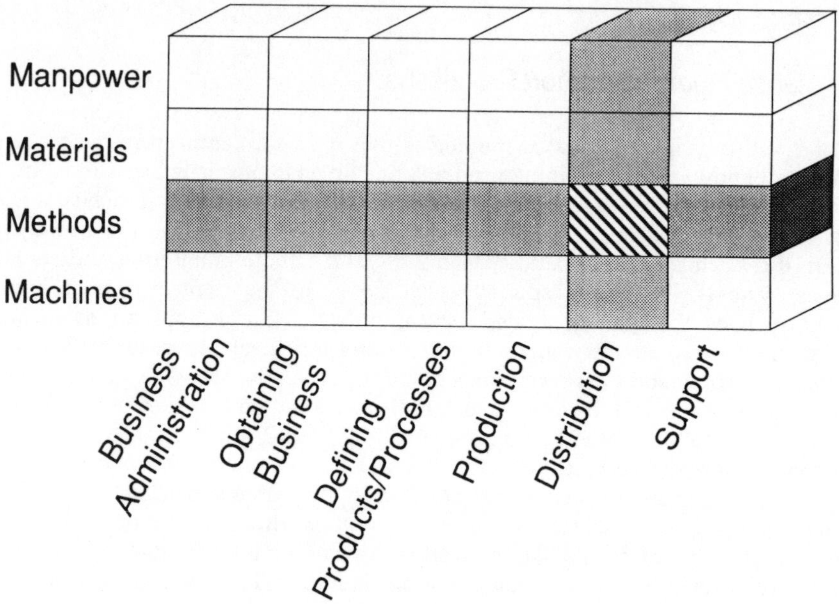

1. Direct distribution to industrial user. Examples include commercial aircraft manufacturers and industrial machine manufacturers.
2. Manufacturer to wholesaler to industrial user. Examples include hardware distributors and industrial safety equipment distributors. About 10 percent of all wholesalers are manufacturer-owned.

The second distribution channel, that for consumer goods, consists of products delivered through the following channels:

1. Direct distribution to consumer. Examples include Goodyear and Firestone Tire dealerships.
2. Manufacturer to retailer to consumer. Examples include Wal-Mart and other retailers who buy directly from manufacturers and sell directly to consumers.
3. Manufacturer to wholesaler to retailer to consumer. Many small-ticket consumer items such as those found in a small drug store or other retail outlets are handled in this manner.

Often, mainstream American manufacturers have utilized more than one of these channels simultaneously. For example, automotive manufacturers such as Ford and General Motors sell vehicles directly to the U.S. government and to major rental car companies such as Hertz and Avis. At the same time, they maintain dealerships to handle single-unit sales to end consumers.

Several powerful factors are at work that make the future of these different methods pretty cloudy.

The Changing Role of the Middleman

One factor that will push some manufacturers away from channels oriented to intermediaries (wholesalers/distributors) is the recent focus on total quality management (TQM). Many manufacturers have discovered—and others have rediscovered—that each additional step in the distribution chain adds another opportunity for the quality of products and services to be compromised. Like any process, total efficiency and quality are the (mathematical) product of all of the process elements. For example, if the manufacturer's quality level is 98 percent, the handled and distributed quality of the wholesaler is 95 percent, and the handled and distributed quality of the retailer is 95 percent, then the quality level of the products delivered to the end consumer will be only 88 percent ($.98 \times .95 \times .95 = .88$)!

Therefore, many manufacturers will be sorely tempted to eliminate the "middlemen" in these processes whenever possible to improve the delivered quality of products to the consumer through more direct control.

Another factor that will influence these decisions is the increasingly global nature of the end-item market. Pressure will continue to mount from foreign countries to offer up larger percentages of the work share in manufactured goods in order to improve trade opportunities in those markets. This will result in even more American manufacturers locating satellite factories in the foreign markets and simply distributing from those market-based production centers. Where wholesale/distribution centers are still required, the ones utilized will almost always be the channels already indigenous to that market (to improve overall market penetration).

Other factors at work in the wholesaler/distributor/retailer scenario are the increasing levels of product distinctiveness and customization as we move into the last half of the 1990s. In some cases, American manufacturers have used distribution channels as a means of achieving product distinction, or setting their products apart. (Izod shirts was one example of this approach. Others include high-end "catalog sales only" product lines.) There is sometimes a perceived marketing advantage associated with "exclusivity" in this scenario, which is sometimes real and sometimes merely imagined. During the next decade, distribution professionals will feel increasing pressure from consumers' desire for "customized" products, which are, if not specially tailored to the individual customer, then at least designed to fill the requirements of unique and smaller market niches. This pressure will have substantial consequences for those who choose to market solely by catalog. Catalogs today can be tailored to any number of unique markets, from book buyers with avid interest in mosses and lichens to consumers of packaged gourmet foods to T-shirt buyers. "Form" catalogs (similar to form letters generated on a word processor or computer) are increasingly used to custom-design the same product for different groups, for example by simply changing the name and logo in the catalog copy from those of one school to those of another or from one family's name and family crest to another's.

As technology advances, this unique hybrid of "targeted" and "mass" marketing will be done electronically, and eventually in virtual reality. Today most distribution operations haven't "cracked" this "code." As a result, wholesaler and retailer outlets are often in a much better position. They offer a broad spectrum of products suiting more diverse customer needs from the product lines of many manufacturers. It seems certain that over the next decade or two, that advantage will disappear.

Another element that will have a significant effect on distribution methods is product complexity. Products are generally becoming more complicated and will continue to grow in complexity. Typically, the more complex a product is, the more likely it is to be sold directly by the manufacturer to the consumer. Countering this trend in the years ahead will be advances allowing the user to set up or install new products, perform self-diagnostics, and make simple operating decisions (e.g., thermostatic controls and light sensors to dim readouts). Consider, for example, how easy it is to set up a desktop computer. It wasn't many years ago that the average worker would never have been able to accomplish such a feat with a device that was far less capable and complex. These days, extremely powerful computers are, as the MIS people are fond of saying, "plug-and-play." When products become this easy to install, use, and service, far less support is needed from the end distributor.

All of these factors will need to be considered as new product distribution strategies are developed for the next decade. They will be incorporated with the traditional objectives of minimizing transportation costs and ensuring adequate transportation networks for enhanced levels of responsiveness to customers. These decisions have generally been made by balancing distribution cost and market share considerations. With the market increasingly fragmented and widely distributed, it seems that the wholesaler/distributor/retailer outlet should benefit, because manufacturers will expect to pay more to distribute a wider variety of goods to a more diverse market. (This is not necessarily because more goods will be consumed; rather, with the added diversity of customer needs will come increased uncertainty about the precise product configuration that will sell. Here again, the increasing importance of merging marketing and distribution functions becomes evident.)

The development of this entire strategy for each new product line will become the domain of those distribution professionals comprising the concurrent engineering team.

Building Integrated Distribution Systems

The other half of the future distribution process will be the deployment of tactics that support the distribution strategy. It is likely that the preponderance of activities associated with these tactics (route planning, fleet management, and management and deployment of finished goods inventory) will continue to be managed through a system or series of systems emulating today's distribution requirements planning (DRP). However, the application of expert systems will dramatically reduce the level of direct human labor required in this field. Most routine decision making related to route planning, triggering of preventive maintenance of distribution vehicles, and decisions pertaining to lot sizes, replenishment levels, and physical deployment of finished goods will be fairly easy to build into expert systems.

It should be a real horse race in terms of who develops these systems first—the manufacturing/distribution professionals or the distribution/service professionals (e.g., Federal Express), who will vie for these services as subcontractors to the manufacturing organizations. The individual manufacturers should understand the product line, market, and applications better. However, the distributors unquestionably understand distribution better.

It will probably come down to who will most effectively integrate with marketing professionals and their information. Because of the potential of colocating with marketing professionals in the concurrent engineering teams, the manufacturer's own distribution professionals are in a better position to win this race. However, it will require the manufacturer to commit resources to developing this software, and many mainstream American manufacturers have not yet recognized the importance of that effort.

Other trends already in place that will grow and continue to affect the actual execution of distribution tactics include the following:

- *Robotics,* which are associated primarily with material handling and packaging at the factory and handling at warehouse facilities in the form of automatic storage and retrieval systems (AS/RSs)
- *Automatic identification systems,* which typically include bar-coding and remote-scanning methods
- *Electronic data interchange (EDI),* which will continue to move toward "seamless" information exchange between the manufacturer and the consumer

In all three of these cases, the single most enabling item will be standards. Companies need to adopt a single communication protocol among all of the systems and equipment in the chain between manufacturer and consumer. This will be one of the most important enhancements to overall distribution system efficiency that a manufacturer can make over the next two decades. Because these standards are still in development, nailing this down will not really be possible in most cases until later in this decade. Norman Weizer of Arthur D. Little describes this situation in his book *The Arthur D. Little Forecast on Information Technology and Productivity.* He points out that three principal network standards—SNA (systems network architecture), OSI (open systems interconnection), and TCP/IP (transmission control protocol/internet protocol)—currently provide the main avenues from which interoperable communications networks may be developed from multivendor systems. All of them are still evolving, and the distinctions between them are disappearing as a result of "gateway" products that connect one with another.

Weizer further indicates that market penetration of these vendor-independent standards will be a function of political, rather than technical, issues. The emphasis of standards development has migrated to a large extent toward the applications level of the OSI model. Weizer predicts that several important OSI-based products could be entrenched in the marketplace by the mid-1990s.

For a while, then, it appears that distribution professionals will face the decision to continue developing and implementing system components at risk, hoping they can eventually be integrated, or else wait for the availability of a mature set of communications standards and forgo the immediate improvements in efficiency associated with (even unintegrated) automation.

Getting Our Goods to Market

The transportation needs of American manufacturers will evolve to reflect distribution strategies and specific tactics selected to support those strategies. The major transportation methods as we enter the mid-1990s are as follows:

- *Water carriers* inland or barge lines and deepwater, oceangoing vessels, which generally provide an inexpensive but slow method of transporting goods. The increasing need for flexibility and responsiveness will limit this mode of transportation as a means of moving finished goods. However, a globalizing supplier network may enhance its suitability for transporting raw materials.

- *Railroads* currently carry more freight than any other mode of transportation. They are best suited for moving bulk loads over long distances. They are experiencing stiff competition from motor carriers and have recently responded with new services such as "piggyback" service (shipping fully loaded truck trailers by rail), fast freight (rapid transport of perishables), and in-transit privileges (allowing transporters to make stops en route to perform functions such as intermediate assembly operations before continuing on to their destination). This mode of transportation is likely to continue to support American manufacturers through the end of the current century, largely because of the aforementioned and other recent innovations.

- *Motor carriers* (usually trucks) carry roughly a quarter of all domestic freight. They are currently most efficient for hauling relatively small loads over short distances. They are extremely flexible because they can go just about anywhere there is a road. They have become much less expensive in recent years due to deregulation of the industry. This relative cost improvement has been the pivotal factor in recent decisions by some manufacturers to increase the volume of manufacturer-to-retailer and in some cases manufacturer-to-consumer deliveries. This allows them to bypass or eliminate the wholesaler/distributor elements of their distribution chain and thus improve responsiveness to fluctuating market demands. Because the domestic distribution strategies developed by American manufacturers over the next decade will undoubtedly reflect the desire to improve flexibility and customer responsiveness, the base of motor freight distribution transport seems solid at least through the first decade of the next century.

- *Air carriers* are, of course, the fastest transport method. Therefore, this mode is generally reserved for those shipments where speed outweighs all other factors, including cost. For this reason, air shipment is used for only about 1 percent of all domestic freight. However, in view of the increasing focus on responsiveness and customer service, and in view of the significant inroads of overnight carriers in many of the formerly traditional distribution arenas, it is likely that this percentage will grow dramatically over the next two decades.

Method	*Comparative Strengths*	*Comparative Weaknesses*
Water carriers	Cost Flexibility	Speed Dependability Locations served
Railroads	Locations served Flexibility	Cost Speed Dependability
Motor carriers	Locations served Speed Dependability	Flexibility Cost
Air carriers	Speed	Cost Dependability Flexibility Locations served

The strongest participants, then, in terms of current distribution transportation methods, are likely to remain motor carriers and railroads throughout the 1990s and into the following decade. In terms of market share growth, with the combination of small share currently held and increasing emphasis on speed, air transport is a likely candidate for substantial growth. Globalization of markets and suppliers is likely to increase water transportation as well.

Looking ahead at some of the major changes in distribution methods, we can see that stocking and inventory controls are likely to be influenced by more fickle market demands and are likely to respond by shortening internal cycle times through automation and other efficiency-improving technology rather than maintaining large safety stock levels. There will be shorter distances required to transport products because of decentralized factory locations, and when the distance is significant, there will be heavier reliance on rapid air transport. Finally, strategic decisions related to factory site location will become more important, as will determining levels of product integration.

The Methods Used in Product and Customer Support

The methods used in product and customer support as we enter the next century will evolve away from the myriad relatively unintegrated functional categories described in Chapter 6 (e.g., delivery of repair parts, field service of the product, return of defective products, warranty claim management, production and delivery of technical publications/instructions, and collection and management of field performance and field failure information). In the nearer term, however, as we move toward the close of the 1990s, another category of functional activity—customer training—is becoming an increasingly significant aspect of customer support.

The methods used in product and customer support.

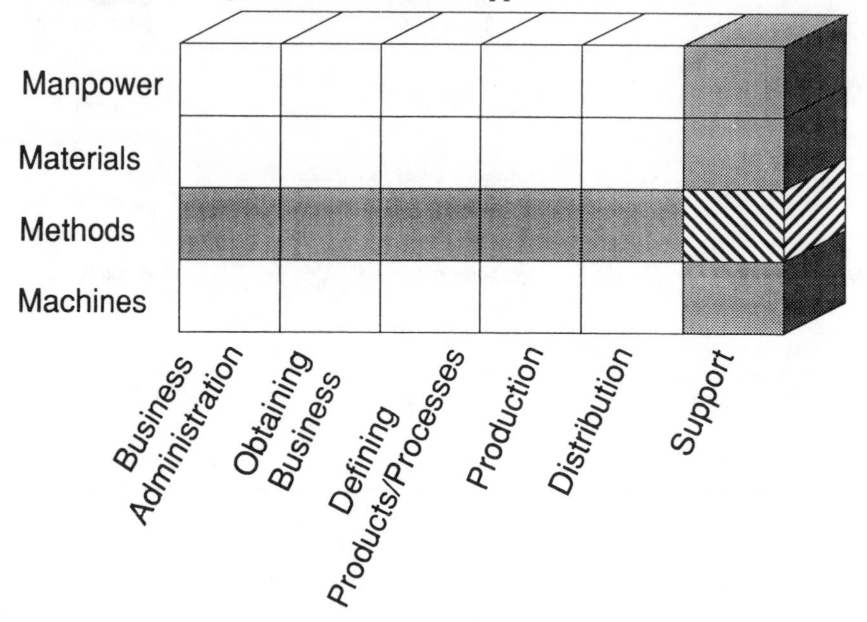

From Customer Support to Customer Training

Although the best-designed products become simpler to use as they become more capable and complex, there will continue to be products that are not so "user friendly," and people who are "just not mechanical." Increased emphasis on customer training is emerging as a by-product of the increased complexity and "knowledge content" of modern products that are *not* "plug-and-play." (One example of the need for this kind of customer support service is the number of people who have not yet figured out how to set the clocks on their video cassette recorders.) Specialized classes have been run for several years by such retail outlets as these:

- Camera shops, which sponsor on-site photography classes
- Computer stores, which offer classes in spreadsheet software, specific computer operating systems, and so forth
- Helicopter manufacturers, which offer pilot and service technician training packages for their products

As products become more "knowledge rich," they sometimes require increasing levels of training for operation. Manufacturers that surmount this requirement as a matter of product design will find that the lack of required training is an incredibly attractive product sales feature. Apple computer, for example, made tremendous inroads in this area with the icon-driven Macintosh, which practically "spoke English" compared to its competitors in terms of being easy to learn to use. The very best designs require the least operator training. A more pervasive example might be the automatic transmission in automobiles, which makes driving a car much faster to learn by eliminating the requirement of simultaneously negotiating a clutch, accelerator, and stick shift.

As we head toward faster and faster introduction of leading-edge technologies, manufacturing companies' customer support organizations are likely to find the area of customer training to be a fertile one, especially in the most complex and "knowledge rich" product lines.

Customer Support as Team Players

As we move into the next century, product and customer support organizations will deploy their people in two areas: concurrent engineering teams and functional organizations.

Members of the concurrent engineering teams will concentrate on incorporating customer needs in product design and on keeping the need for product maintainability a high priority in design/cost trade-off decisions. Specific activities performed by product and customer support people on the concurrent engineering team will include the following:

- Surveys of product users to identify their needs and incorporate them into the quality function deployment (QFD) process
- Oversight and coordination of customer support advisory teams to identify customer problems and feed them back to the concurrent engineering team as they arise
- New design review participation to ensure product maintainability and supportability
- Participation in design change review boards to ensure product maintainability and supportability
- Drawing review to ensure usability of drawings for technical publications, owners manuals, and service/repair publications

Specific activities of the functional product and customer support organization will include the following:

- Service center management
- Spare parts requirements forecasting and administration
- Administration of repair, overhaul, and exchange activities
- Tracking of product performance in the field, especially the first products put in use, to limit the effect of any unforeseen problems immediately upon their discovery
- Customer training
- Management of technical support and services, including those offered by the manufacturer as well as those offered by subcontractors
- Warranty administration
- Management of "customer response centers"
- Infrastructure support administration

Beyond Product Support: Data Bases and Infrastructure

There are at least two significant differences in product support and customer service that will continue to become more distinct as we pass the year 2000: the data base orientation of customer and product support management and the importance of infrastructure support.

Virtually all customer and product support will be founded on a network of data bases by the end of this decade. Individual data bases will be maintained at point-of-sale and point-of-service, where such records as service and operations reports, service contract reports, customer address and other demographic records, publication updates, and warranty and repair input records will be maintained. All of this data will be fed to a set of aggregated data bases that will be held at service center or manufacturer locations. They will be used constantly by a wide range of manufacturing organizations for everything from marketing to distribution, as well as for the traditional customer service applications. (See Exhibit 12-7.)

The other major change in product and customer support methods through the end of the 1990s and in the following decade will be an increased emphasis on infrastructure development and support. "Infrastructure" in this context means the physical, legislative, and cultural environment of the product and the customer. By concentrating on the infrastructure, product and customer support organizations will move from their current (largely) reactive mode of operation to a more proactive one.

One to two decades ago, most of the activity and methods of the support organization were focused on product support. It was simply a matter of maintaining an adequate supply of service manpower, tools, and repair parts on hand at enough locations. Then, in the late 1970s and 1980s, the customer became more widely recognized as an appropriate object of support. Techniques like QFD and total quality management stressed customer satisfaction as the road to success. (This proved difficult and still poses significant challenges for manufacturers today. Correctly identifying the needs of customers is a far trickier proposition than it would first appear, and is a much less scientifically definable process than identifying technical problems with the product.) Now, as we approach the second half of the 1990s, product and customer support organizations are beginning to recognize that the environment of the product and its user is extremely important as well. As products grow in complexity and value, managing infrastructure is becoming a cost-effective means of ensuring product line success.

Elements of the product and customer infrastructure include the following:

- *Product usage platforms.* For example, in the airline industry this would include adequately sized and configured runways and other airport facilities.

- *Legislation.* This area has been recognized as vital in some industries for some time, and is actively pursued by a few, including automotive manufacturers dealing with issues such as Environmental Protection Agency (EPA) regulations. Special interest groups, political action committees (PACs), and lobbyists are tools already widely employed by some.

- *Product-related service industries.* Many more manufacturers will learn to develop support strategies that utilize cost-effective, independent service busi-

Exhibit 12-7. The product and customer support process.

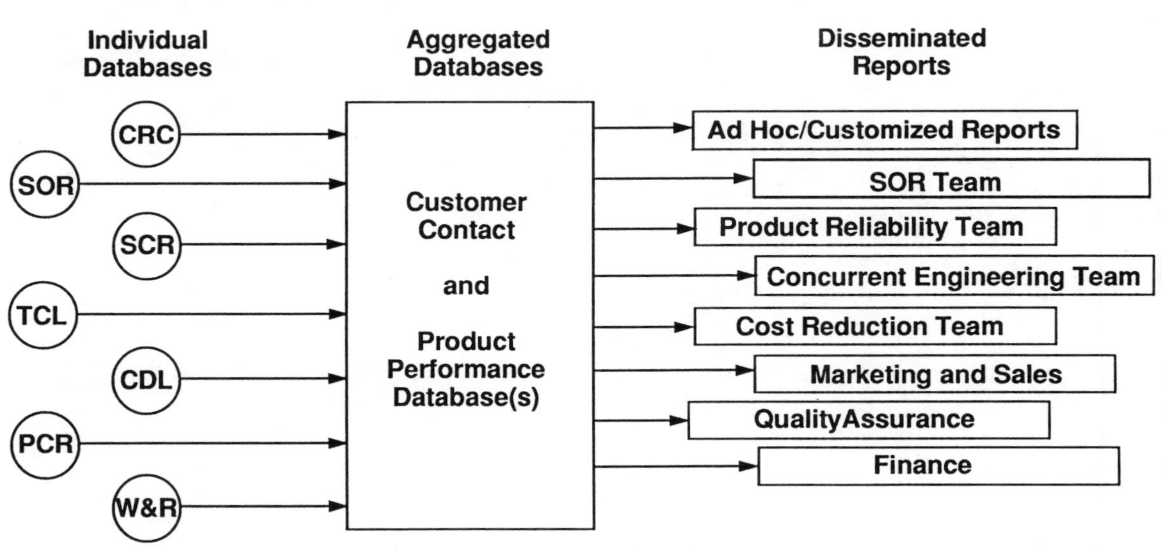

nesses to provide product and customer support services such as financing, parts warehousing, product upgrades, overhauls, and exchanges.

• *Technical innovation incorporation.* The competitive edge as manufacturers become more focused on cycle-to-market will be held in many cases by the manufacturer that most rapidly identifies and incorporates technical innovations identified in the marketplace. Customers often not only identify potential product improvements (e.g., lengthen the handle, add an automatic feeder, and so on) but also incorporate them in products after they buy them. Instead of just admonishing customers that modifying the product will void its warranty, product and customer support organizations will recognize potential product differentiators and incorporate them as either standard features or optional modification kits.

• *Cultivation of user groups.* Where user groups already exist, such as personal computer clubs or Chevy owner's clubs, the industries who already have these folks as customers, and those who wish to have them as customers, will learn to methodically identify and infiltrate the groups to learn everything they can about user needs and preferences. In situations where groups do not exist, manufacturers in some industries will start them and sponsor them over sustained periods. (Some software companies have already done this.) Trade associations and other professional groups, as well as nonprofit organizations and public interest groups, will also prove to be fertile ground in these endeavors.

Summary

The dominant methods utilized in American manufacturing organizations will be diverse and dynamic over the next twenty years. Yet they will have a number of common "threads" woven throughout. In almost every area, we will see teams at work. They will usually be interdisciplinary teams, as best typified by the concurrent engineering approach. People (again, usually teams of people), will be more "empowered." With flatter organizations and dramatically improved multimedia communications, spiderwebs of teams will form incredible electronic webs that support multiplant, multiregional, and even multinational manufacturing management. Expert systems will give way to more sophisticated knowledge-based systems, and hence continue to reduce what we have traditionally known as "touch labor" work methods. Functional silos will blur, as the best of our manufacturing companies convert to process-based management and improvement. Eventually, even the lines between supplier, manufacturer, and customer will blur as suppliers and customers become an intricate part of our product (and therefore process) definition activity. Again this will be facilitated and accelerated by multimedia telecommunications, as well as by some of the economic influences discussed in earlier chapters in regard to international supplier bases developed to improve sales opportunities in the national markets of those suppliers. Finally, in general, both our products and our processes will themselves become more and more "intelligent." As a result, the best manufacturing companies will become true "learning" organizations, developing their processes and associated process-oriented training and education as a part of ongoing production.

Chapter 13

Dominant Materials: From Software and Silicon to Composites and Ceramics

When you read the answer to this question, you'll probably think that it was a trick question. It isn't. And the answer is quite profound. The question:

> *"What will be the single most important material used by manufacturers in the coming decades?"*

Think about that. Certainly, there will be a host of new and exotic materials developed, introduced, and implemented over the next twenty or thirty years. But which will be the *most* important? Composites? Ceramics? Superalloys of ferrous or nonferrous metals? Some new polymer perhaps? No. The single most important material used by manufacturers in the future will be data. The ability of a manufacturer to gather, assimilate, analyze, and convert data into useful information will be the single most important determiner of success. Market data, process capability data, material characteristics data, field performance data, and other types of data will all be the raw materials of the future, and how manufacturers handle this precious resource will distinguish the winners from the losers.

Just as cycle-to-market is critical in today's manufacturing environment, the cycle time of gathering, converting, and utilizing data throughout the company will be a vital, determining factor in future manufacturing competition. Therefore, when you notice that a great deal of the "materials" listed in the balance of this chapter are in fact data, don't be surprised. The real surprise is that it has taken so many of us so long to recognize this resource for what it is.

The physical materials will, of course, evolve as well. In the production world, we will see a continued shift from metals toward composites, ceramics, fiber optics, and other alloys of these materials (see Exhibit 13-1). Beyond these changes, we will see the materials shift to a more elemental level. At some point many years down the road, the raw materials will be chemical "slurries" and raw elements that may be disassembled and reassembled or restructured in replication processes. These are discussed in more detail in later sections of this chapter.

The Materials Used in Business Administration

The materials (primarily data) used in business administration in twenty-first-century manufacturing companies will be much different from those used today

Exhibit 13-1. Dominant U.S. manufacturing materials.

U.S. Manufacturing Materials

Business Administration	Obtaining Business	Define Products and Processes	Production	Distribution	Support
• Different data • Time-based management data • More customer intelligence • Environmental data • Strategic initiatives status data	• Heavily focused on "customer needs" data • Simulated test marketing	• Assembly and component models • Performance data • GT data • Materials capability data • Standard data • QA strategy • Product/process cost methodology • Knowledge-based systems software • Process capability data • Marketing information • Packaging requirements • DFL data • Product performance specs • Procurement strategy • Supplier process capability data • Customer requirements data	• Schedule data • Inventory data • Bills of material/bills of resources • QA data • Procurement data • Training materials • Process knowledge/intelligence • Simulation software • Constraint management software • Materials development software • Metals/alloys • Ceramics • Paper • Wood • Plastics • Composites • Fiber • Stone • Transformation-toughened ceramics • Fiber-reinforced ceramics • Photo-sensitive polymers • Magnetic materials • Biologically derived materials • Routing data • Labor/time standard • Manufacturing methods data • Maintenance data • Cost data • Configuration data • Critical characteristics data • Decision support software • Process management software • "Teaching factory" software • Inter-metallic compounds • Single-phase metals • Metal-matrix composites • Microduplex alloys • Radiation-resistant alloys • Micro-cracked ceramics • Piezoelectric substances • Amorphous silicon • Optical memory material	• Delivery schedules • Route data • Third party data • Inventory data • Bar code/ID data • Advanced containerization materials, cushioning materials, barrier materials, shielding materials • Market/sales forecasts • Cost data • Field performance data • Delivery performance data • Packaging material data	• User survey data • Benchmarking data • Customer advisory team feedback • New design data • Service center management data • Customer training data • Technical support/service data • Infrastructure data

The materials used in business administration.

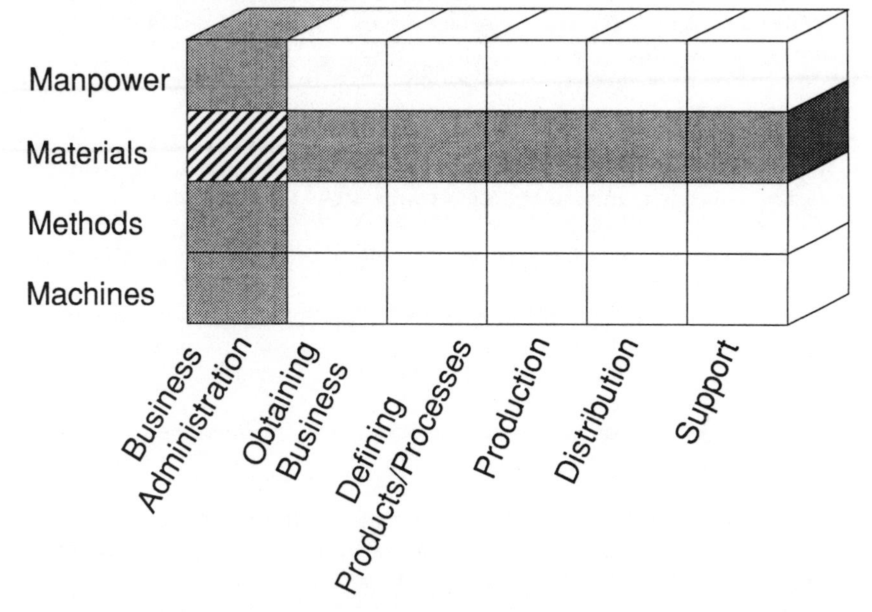

in several respects. The data will be different because business administration professionals will manage differently, and therefore value different kinds of data. The data will be handled differently because more of it will be electronically maintained and held rather than stored on paper. It will be aggregated differently because performance measures will be different. It will be maintained and analyzed at different levels, reflecting changes in the way the company is organized.

Specifications for the "Postmodern" Factory

In a *Harvard Business Review* article (May–June 1990), Peter Drucker says, "We cannot build it yet, but already we can specify the 'Postmodern' factory of 1999."

He goes on to describe an environment where the primary unit of measure is time, where statistical quality control, new manufacturing accounting methods, and modular manufacturing are dominant, and where management is typified by the "systems approach." These changes will entail alignment of information with accountability to more effectively manage the enterprise and will require constant and reliable feedback from every aspect of operations. It will involve collecting data that measures nontraditional costs, moving the emphasis away from direct labor measures and toward a material and activity basis. The factory will be far more easily recognized as an information network than as a physical entity because of the bridges built by telecommuting and other forms of EDI.

Business administration professionals will tend to measure things such as "mean time to respond to changing customer demand," focusing on optimizing the velocity with which value is added to the product, from order entry through delivery and field support. More important still, top management will require constant information that demonstrates that the needs of the customer are accurately identified and consistently satisfied throughout the organization. They will demand constant proof that all major processes are in (statistical) control and that the company is meeting or exceeding "best in class" levels in important performance areas.

Top executives will be compelled to develop and maintain a crystal clear long-range vision of what they wish to be and monitor their progress toward achieving that vision. This will require information pertaining to the following:

- The image of the company as it is perceived in the marketplace and the community
- The values of the company as perceived by customers, the business community, and the employees
- The products of the company and the degree to which they meet existing and anticipated customer needs
- The performance of the company as it relates to the financial, market share, and other strategic incremental targets

Managers will also need to monitor performance data and environmental data pertaining to critical business issues such as new competitors and competing products, comparative product quality levels as perceived by the marketplace, responsiveness of the company to market changes, and comparative data about the market's perception of product value.

Finally, top executives will require at least summary level data reflecting the status of major strategic initiatives, programs, and projects. Management will be more intensely focused than ever before on resource allocation information in this area, especially the allocation of human resources. Because of recent and future increased emphasis on the importance of our human resources assets, management will be interested in tracking those aspects of company operations that influence employee morale and motivation. For example, regular evaluation and analysis is likely to be performed to measure employees' perception of these aspects:

- Level of challenge in work performed
- Whether their opinions are valued
- Whether their best work is recognized
- Whether their pay is clearly tied to their performance
- Whether their performance is fairly evaluated
- Competitiveness of employee salaries (perceived and real)
- Clarity of performance goals
- Their career paths
- Employee harmony
- Competitiveness of employee benefits (perceived and real)

In many cases, we don't even know how to measure these elements today. Yet measuring these and other factors like them will be a necessary part of measuring the quality of twenty-first-century human resources management.

There will be a concerted and deliberate shift toward life cycle costing in companies not already on that path, which will also mean that much different data aggregation will be required. (These efforts are likely to track closely with the company's progress in concurrent engineering efforts since a primary function of concurrent engineering is to shift product costs from the production phases of the program further and further upstream.)

Different levels of integration will occur over time, making this information "material" more or less difficult to manage. As we move toward the end of the decade, this information will move from unrelated paper files or computer files through a level of integration characterized by electronic mail, then through a level of common contexts and format/content uniformity, to a level of common or fully integrated data bases. Obviously, the further along the company is in this evolution, the easier it will be to gather and manage the appropriate information for future business administration activities.

Beyond mere integration, information-level linkage will be an extremely important aspect of the information "materials" used in business administration. This can best be visualized as a structured bill of material, with pieces of information instead of parts and assemblies. Constructing the information aggregation architecture required to support meaningful performance measurement in the areas described above will be no small task. However, great strides have already been made in this vein by companies like Digital Equipment Corporation, under the auspices of their Computer Integrated Enterprise effort.

The Materials Used in Obtaining Business

Obtaining business will become an even more interesting activity as we move to the close of the 1990s than it is today. As discussed in Chapter 12, the methods

The materials used in obtaining business.

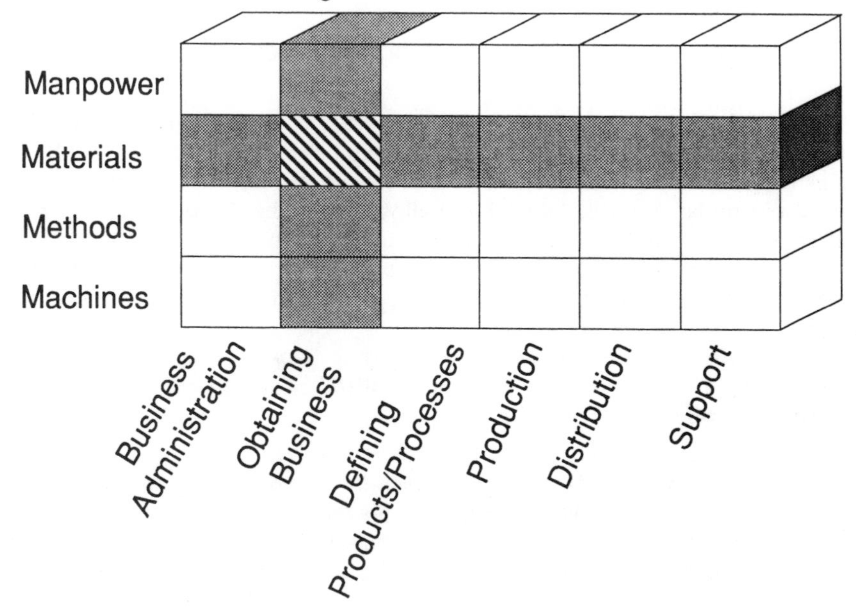

and tools of "marketeers" are moving away from the traditional focus of How do I sell more of this product we offer? and more toward How can we make money by satisfying the needs of our (existing and potential) customers? Because of this shift, the information "material" of the marketing and sales professionals will increasingly consist of "customer needs" data.

Although salespeople will continue to be interested in historical sales data, it will become less and less prominent in their daily work. Product life cycles will continue to shorten, and cycle-to-market will continue to be hammered. For example, consider the following recent observations:

- In electronics, the first two companies to get new-generation products out to the marketplace typically capture 80 percent of that market.
- Since 1989, we have moved from an environment where 45 percent of the products sold were introduced within the previous two years to an environment where (by 1992) over 60 percent of the products sold were introduced within the prior two years.

While this certainly doesn't obviate the value of repeat buyers, it clearly indicates that more marketing and sales time and money will be best allocated to identifying and satisfying customer needs. If we don't, our competitors (both foreign and domestic) will. We will be able to see things as Sony did when they determined that both existing customers of tape recorders/players and noncustomers alike had an unfulfilled need for very small, portable, belt-mounted tape player systems. Every dollar spent on that kind of marketing and sales activity will surpass hundreds of dollars in personal contacts to sell existing (but no longer novel) products. The kind of informational material used, then, will reflect this change in the sales environment.

Testing the Market Without Leaving the Office

One of the most significant and visible aspects of this focus on identifying and satisfying customer needs will be the use of simulated test marketing (STM). As mainstream American manufacturers become more comfortable and proficient in the use of simulation technology (almost all will be by the end of the 1990s), STM will become the central method for performing consumer market identification, establishing brand awareness strategies, conducting (simulated) brand trials, projecting repeat purchase volumes, estimating sales, and establishing likely profit targets.

STM is described in detail in an outstanding book recently written on this topic by Kevin Clancy and Robert Shulman entitled *The Marketing Revolution*.

They describe STM as follows:

> A simulated test market (STM), in which the computer program represents the national market, gives us answers quicker (three to five months versus twelve to eighteen) and less expensively ($100,000 versus $3 million) than a "real-world," in-market test.

In simulation, this system introduces a product (or service) to the target audience, measures market response, and projects the outcomes of the product through completion of its full-scale launch.

In the current environment, the forecasting systems and approaches used by marketing and sales organizations utilize primarily "year-end awareness" values, "awareness-to-trial" values, distribution estimates, forecast penetration estimates, repeat purchase values, and overall usage data to forecast sales. However, many of these values and estimates are built from fairly spurious data and in a somewhat haphazard manner.

STM systems capture virtually every vital element in the marketing equation. These systems then test any and all of the plans the marketing and sales professionals care to evaluate. They allow managers to ask what-if questions to make trade-offs, such as, What area of my marketing cost could I reduce and have the least impact on sales? or Where would an increase of 10 percent in spending yield the greatest benefit? and so on.

Much more extensive data is required to run such a system than the amount utilized by current systems, and it requires constant updating to retain its value. STM systems use the following:

- Media weighting factors, media impact assessment values, and advertising impact assessment values to identify the size and structure of the consumer market that will be created.
- All of the data used in the first item in this list, plus distribution data and couponing and sampling data to evaluate and identify levels of brand awareness.
- Distribution, couponing, and sampling data, product concept information, advertising persuasion values, and packaging impact values in conjunction with brand awareness level estimates to project initial purchase levels.
- Pricing, customer satisfaction levels, and product-life-level data along with the initial purchase levels to assess the quantity and distribution of repeat purchases.
- The purchase volume data calculated in no. 4 is then converted through unit price, discounting schedules, and so forth, into sales dollar volumes.
- Finally, applying marketing, production, and distribution costs yields anticipated profit levels for each product line by year, by region, by demographic feature, and so on.

The cascading nature of these simulations requires maximum data integrity at each step, because error factors are multiplied as the next step is performed. The volume of data, accuracy of the data, and timeliness of the data are as vital to the marketing analysis as any raw material is to the production process. In fact, there are real parallels between the two:

Material Features

Marketing	*Production*
1. Bad data early in process makes all	1. Bad material causes quality of parts

subsequent analysis suspect or in some cases worthless.

2. Data must be maintained continuously or it becomes less and less likely to yield valid estimates, and therefore results in poor decisions.
3. Guessing about the data will cause expensive downstream work in marketing, production, and distribution costs.

and finished products to be reduced, resulting in rework or scrap.

2. Raw material often has a shelf life, and when used beyond that threshold will often result in defective products.
3. Poor material use in production will cause expensive overruns in material and therefore cost.

Because of the continuous updating and validating of data that is required for maximum accuracy, many manufacturing organizations will elect to buy their STM as a service from marketing and consulting firms. Others will attempt to maintain their own data bases, utilizing the vast resources of on-line data bases that are becoming available even now. These companies will still be required to gather a great deal of information themselves, but some will be in an excellent position to do so because of the data management systems already being constructed in distribution and point-of-sale locations.

The Materials Used in Defining Products/Processes

As we close out the twentieth century, the materials used in defining products and processes will continue to consist of the same two major elements we use today: data and software. However, the data used will be more voluminous with the advent of concurrent engineering, and the software will be more complex, owing to the advent of knowledge-based systems.

Most of the data we use today to define products and processes will be relevant for another twenty to thirty years, until the advent of nanotechnology eliminates the need for macrolevel material processing in production as we know it today. (Even then, production process capabilities will be an important factor. However, they will be determined more by the materials' molecular composition and chemical stability/predictability than by the tolerances and repeatability of specific pieces of fabrication or assembly equipment.)

Dependent on Data

Over the next twenty years or so, then, we will continue to require accurate and timely data in the following areas:

- Assembly and component models, the three-dimensional electronic images that are replacing two-dimensional drawings. A dramatic change in this area will be the electronic movement of these models for multiple purposes. For example, the models will be available on shop floor terminals, eliminating obsolete drawings

The materials used in defining products/processes.

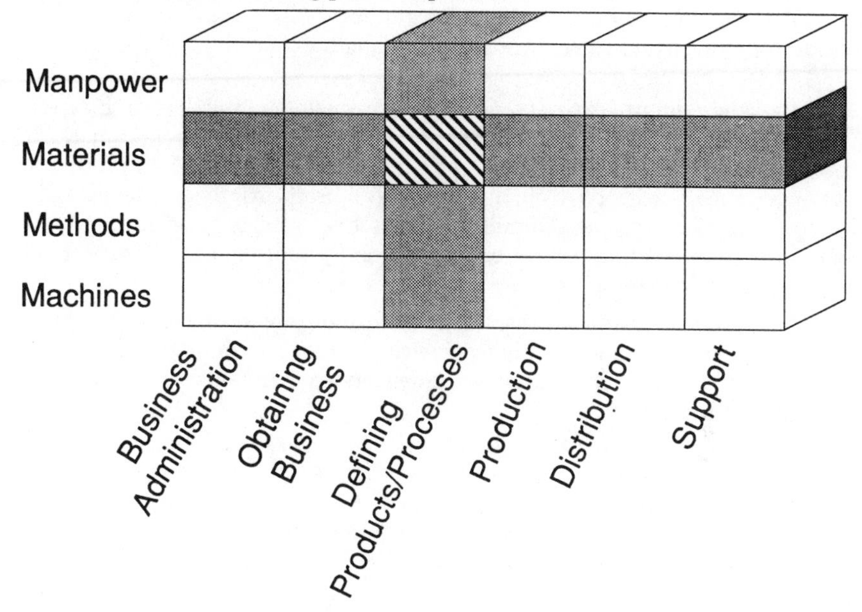

caused by "rev level" changes (significant changes in the design of a part), providing up-to-the-minute quality assurance data (e.g., inspection-related dimensions, tolerances), and offering advantages in terms of training for fabrication and assembly personnel. In addition, some leading-edge manufacturers are already sending these models via EDI to suppliers, who download them to CNC (computer numerical control) equipment and make parts directly from the models. (This process is euphemistically referred to as going from "art to part."

- Performance data, based on field and test lab reports. Because of recent advances in EDI, this data will continue to be acquired faster, and it will be more accurate as a result of fewer intermediate "translations" of the data between the point of origin and the engineer.

- Group technology data, which will grow in importance as the movement toward focus factory operations ("factories within factories," which produce some level of components or subassemblies complete from start to finish) and just-in-time fabrication processes grows.

- Materials capability data, especially properties of generic materials as new materials and applications become prolific.

- Standard parts data, which will also be increasingly important as product standardization grows in conjunction with JIT production applications.

- Machine (and limited process) capability data, which will move from our current (infrequently updated) listings of generally advertised machine capabilities to demonstrated capability and process repeatability data automatically fed back from the shop floor devices themselves.

In addition to these areas, which are the mainstay of current product and process definition, new data will be utilized in the definition process through the development of concurrent engineering, including the following:

• Marketing information, particularly information pertaining to marketing strategy. It will be important for designers to know and incorporate marketing strategy in many areas, for example, whether to design a high-tech product for upper-end markets with relatively high unit-level profit margins, or focus on low unit cost and high volume. Durability versus disposability, operating environments, and several other strategically important elements will be vital to the concurrent development of products and processes.

• Packaging requirements, especially with the growth of rapid distribution and delivery, will become a significant element of concurrent product design. Product modularity and configuration-for-delivery requirements will substantively affect many product designs.

• Design for logistics (DFL) data, closely tied to packaging requirements, will also be increasingly important. Some of this emphasis will result from the globalizing marketplace. However, even in the domestic marketplace, design decisions will be affected by the level of configured product handled and moved. For example, it will become more important to make effective decisions about whether to distribute a fully configured product, or send out a base product that is later configured at distribution sites or points of sale.

• Engineering data, including performance specifications, physical envelope dimensions, weight requirements, hardness, and so on. In the case of polymers and other chemical products, chemical and physical properties as well as shelf life requirements will continue to be specified.

• Production process design data. This data will grow out of the current parts classification and coding (C&C) aspect of group technology (GT) as it converges with cellular manufacturing and flexible manufacturing systems (FMS) to produce generic process routings with defined and repeatable "total process" capabilities. It will allow process designers to apply current total process capability constraints during product design and will support the development of new processes where appropriate.

• Quality assurance (QA) strategy and approach information. As production methods and processes are defined by the concurrent engineering team, it will become quite easy to develop appropriate quality assurance strategies and tactics as well. When the parts will be produced on an NC mill, for example, the mill and tool combined with electronic sensors may be utilized as an automated medium of inspection. (The problem existing in most manufacturing environments today is that, at the time the part is designed, it is unknown exactly how or on what equipment the part will be produced in most cases. Therefore, imposing a specific medium or method of ensuring quality is difficult, and often prohibitively expensive.) Developing the QA strategy and tactical applications will provide a far superior overall design package and a better-quality, lower-cost product.

• Procurement strategy and vendor capability data will become extremely important as the supplier base continues to become a larger part of most manufacturers' production strategies. It will be possible to evaluate in-house capability and compare it to available outside sources of supply while the product is still in design, thereby selecting the most appropriate work to retain. This, in turn, will allow us to maximize our own capacities.

It will also allow us to bring suppliers into the concurrent design process relatively early (certainly much earlier than we typically involve suppliers today) in many cases.

- Process and product cost methodologies will change over the next several years, moving increasingly toward activity-based costing. As they do, the primary factors previously considered in the cost equation will change in prominence. These changes will affect make/buy decisions, especially as overheads are allocated in different ways. In addition, costs often overlooked entirely at this stage, such as equipment reliability and maintenance factors, will be rigorously and far more accurately applied.
- Customer requirements information will be gleaned, analyzed, and incorporated in unprecedented volume as tools like quality function deployment (QFD), benchmarking, and customer "blue ribbon" panels become the rule rather than the exception for American manufacturers.
- Information pertaining to the field support of the product and the customer will play an important role in the concurrent design work of the future. Product-resident interface receptacles for repair diagnostic equipment will be required, self-diagnostic equipment will be designed and incorporated into the products themselves, and in some cases self-repairing products will be designed.

The diversity and volume of information that needs to be incorporated in product and process design continues to increase exponentially. It is among the most vital information in the manufacturing organization, and will grow in importance as concurrent engineering and process-based management become the central focus of manufacturing management.

Expert Systems Need Expert Software

The software "materials" used to define products and processes over the next two decades will change as dramatically as the information that is used. The primary improvement in this area through the current decade and probably well into the following one is likely to be in knowledge-based systems.

Some knowledge-based systems are already used in steam turbine generators, diesel locomotives, and wave-soldering machines. In these applications, the knowledge of specific technology subject experts is captured, reproduced, and utilized to control an ongoing process. In the future, these systems, applied to product and process development activities, will be enhanced further so that they are capable of "learning" from examples and identifying cause-and-effect relationships.

In 1991, the National Academy of Sciences published an outstanding book entitled *The Competitive Edge, Research Priorities for U.S. Manufacturing*, which states,

> Superior AEMs (advanced engineered materials) are expected to result from the use of expert systems technology to integrate systematically

> processing knowledge and materials science. The success of the reasoning tools needed to integrate symbolic knowledge of materials design, process planning, and IPM (intelligent processing of materials) will depend on the understanding of the underlying materials science and the logic of the problem-solving approach employed.

Not only will these systems support the development of new materials, they will radically improve both product and production process development and definition. In product definition, for example, knowledge-based systems will call in required expert subsystems that deal with aspects such as dynamic modeling, geometric modeling, stress analysis, load analysis, kinematics, and thermal analysis at the appropriate times during design activities. These systems will alert design teams to potential problems at the earliest possible stage of development, and will eventually prevent many of the problems that can only be identified through downstream testing today.

When these functions are performed, a great deal of the risk that exists in new product introductions today will be driven back into earlier stages of the development life cycle, as shown in Exhibit 12-4.

The Materials Used in Production

The materials used in production over the next two decades will consist of three major types: data, software, and physical raw materials for conversion in the production process.

Production will continue to utilize many of the data elements it does today,

The materials used in production.

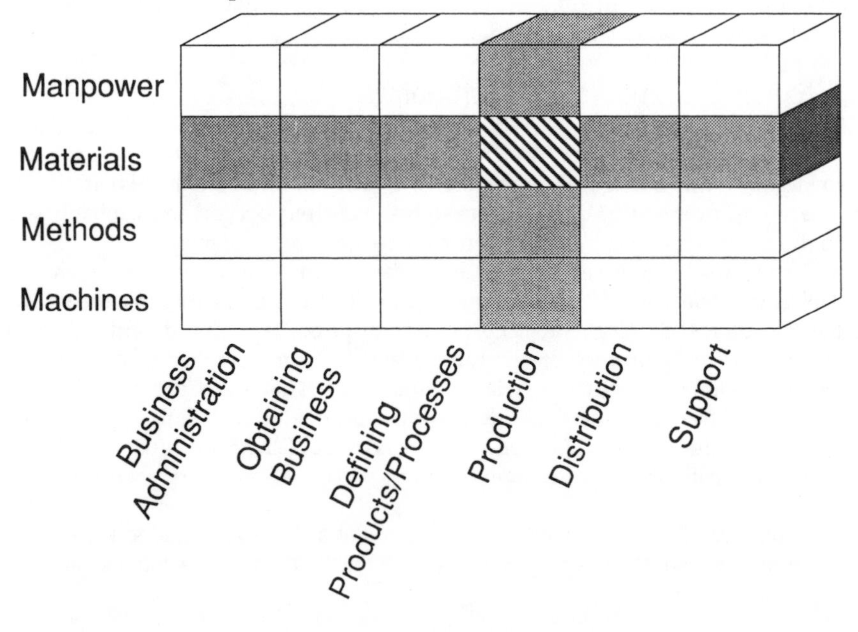

but often there will be changes in the way the data is gathered, aggregated, analyzed, managed, and stored.

New Twists for Old Data . . .

Among the current categories of data that will be retained through the foreseeable future are the following:

- *Schedule data.* The primary differences between the schedule data we use today and schedule data as it will exist at the turn of the century in mainstream American manufacturing operations is that the schedules will be end item–driven, and in most cases will not be carried back in recalculated form to individual machines or work centers as they are today. Individual machine schedules, even in fabrication facilities, will become unnecessary and irrelevant as machines and work centers are absorbed into manufacturing process cells and flexible manufacturing systems. For the most part, work will be "pulled" (as in "pull system" or *kan-ban*) from one manufacturing process cell to another, with no intermediate work center or machine-level scheduling required. The scheduling done will be at the finished product level, and will be "pulled" backward through the production process, even through supplier operations in many cases.
- *Inventory data.* Inventory will still be held at three fundamental levels: raw materials, work-in-process (WIP), and finished goods. However, inventory units of measure are likely to evolve in future manufacturing environments to an increment based on the process—a "process unit" or "pull set" (loosely defined as the quantity of units pulled at one time from process to process). This unit of measure development will reflect the process and cell orientation of manufacturing floor operations and simplify most inventory accounting activities. Especially in environments where inventory is decreased systemically via "backflushed" bills of material, this is likely to be a popular approach.
- *Bills of material (BOMs).* The "process units" or "pull sets" just described will likely change the way inventory units are expressed in bills of material. In addition, bills of material constructed in this manner will more easily support evolution to an expanded version of the BOM known as a bill of resources (BOR).

The BOR will aggregate the total resource requirement for a process unit, including tools, labor, material, and inspection. This, in turn, will be extremely beneficial in capacity analyses, cost estimating, and total resource requirements planning.

- *Routing data.* Routing data will be much more streamlined. Today's production routings still call out, for the most part, individual machines and work centers. Future routings will need to show only a few major processes or cells to cover the same number of conversion steps.
- *Labor and time standards.* The standard unit of time, currently held in manufacturing management systems as a ratio of pieces produced to time required, will change in the new environment as well. In a "pull system"–based production setting, the incremental standards will be process units or pull sets produced per increment of time. Especially in those companies that successfully implement synchronous production, the standards will reflect with exactness the actual

production rate. In these settings, individual process work speed will be tied directly to the speed of delivery of the final product. In these environments, WIP inventory and finished goods will be minimized because the process operators will be constrained by material availability to not exceed the standard rate. The rate will no longer be a way to offer incentives for employees to increase output; it will be relegated to capacity analysis and planning.

• *Manufacturing methods data.* This data, which includes general production work instructions, installation drawings, and so on, is currently paper-based in most manufacturing companies. This causes several problems beyond the mere waste of paper-related natural resources. For example, it means that configuration control of shop floor data must be handled manually and rigorously to ensure that products are built from the correct components and to the proper engineering specifications. As we move toward the end of the decade, computer terminals and other similar display screen devices will be utilized on the factory floor in place of paper. This will eliminate the configuration management problem, provide improved information availability, and offer enhancements such as 3-D views, rotation, and even the animation of drawings simulating fabrication, assembly, and installation processes.

• *Quality assurance data.* This data will evolve from mere specification-based inspection criteria and information used to provide corrective action toward a defect prevention data base. It will evolve through encompassing more data (links to process capabilities, material properties, fail-safe techniques and applications, and trend analyses) and processing/displaying the data in different ways—for example, real-time SPC (statistical process control) run charts, which track and display critical process capability and performance levels.

• *Procurement data.* This area is currently composed in most cases of information pertaining to vendor delivery performance, prices, and product defect rates. In the future, it will also include supplier process capability, supplier capacity and utilization levels, supplier learning curves, and projected delivery performance. As suppliers truly do become partners in the manufacturing process, the information will often be made available on-line directly from the supplier in "real time" via electronic data interchange (EDI).

• *Training materials.* Training materials will change in many mainstream American manufacturing organizations from their current live instructor and paper information basis toward video and diskette or training tools. This evolution will occur throughout the balance of this decade and probably a few years into the next. As knowledge-based systems reach widespread application somewhere around 2005, these systems will begin to be supplanted with teaching systems that gather process data and learn from it in order to teach future operators. (Knowledge-based systems in this setting are discussed in more detail later in this chapter.)

• *Maintenance data.* Maintenance data in use today primarily pertains to machine failures experienced. Some of the better current systems utilize specific information regarding the components that fail in order to identify appropriate preventive maintenance intervals. Future systems will identify far more of the relevant elements related to maintenance requirements, for example, demonstrated lubricant life and expected life of high-wear machine components. They will also include demonstrated effects on useful life of different types of materials, feed rates, and run speeds. Finally, in the latter half of the next decade, handling of this data by knowledge-based systems should provide genuine predictive information supporting breakdown prevention and machine life extension.

• *Cost data.* Cost data associated with production processes will be an increasingly important element of shop floor management as work cells and flexible machining systems (FMSs) are linked to discernible processes and those processes are run like individual profit centers. However, the nature of cost collection will be different, as costs are collected at the cell, FMS, and process levels rather than at the individual operation level. They will also change in nature because of the advent of activity-based costing and because of the movement toward process units/pull sets. Finally, they will change in nature as they are managed in increments of the product life cycle.

The changes related to costing activity (rather than allocations) will result from a need to more accurately define the levels of value added in specific product and process flows. The changes in aggregation levels will occur as management endeavors to more closely reflect actual shop floor management methods. The changes related to product life cycle will become vital as product life cycles become shorter. In our manufacturing environment at the turn of the century, the shortened life cycle of new products and the importance of life cycle element management (i.e., cycle-to-market) will make life cycle costing a primary management method.

. . . And New Data Needs

In addition to these new twists on traditional data, there will be new kinds of data as well. Among the new data types will be the following:

• *Configuration data* for homogeneous production facilities and processes. Production processes in future manufacturing operations will need to be configured to provide optimum levels of capability, capacity utilization, and flexibility to incorporate new product and engineering changes. This is a field we have not yet explored much, because our processes have not typically been modular enough to make use of these analyses. (Six-thousand-ton sheet metal presses buried stories deep in concrete are not good candidates for rapid physical redeployment.) However, our future manufacturing processes will incorporate more "exotic" materials that require less massive equipment, employ more flexible and versatile machinery, and generally have fewer physical constraints. Therefore, a synthesis of process capability, capacity, tooling requirements, manpower requirements, and maintenance requirements will be developed for future production operations.

Synthetic data will be juxtaposed against product design data and production volume requirements in the concurrent engineering process to develop initial process design configurations. Once this data is deployed, collecting and maintaining it will be a function of production operations management on the shop floor. Process reconfiguration may be handled there as well, but might also remain a concurrent engineering function. Responsibility for this work will probably differ from factory to factory.

• *Process knowledge and process intelligence.* Knowledge about process capability will be developed by monitoring specific process elements as they are varied during normal production. Speeds, feed rates, material types, maintenance levels, and even environmental factors such as temperature and humidity will be monitored, recorded, and correlated to the quality and productivity of processes. In this way, the labor-intense work currently performed in designing and executing experiments to solve production process problems will be done continuously and automatically in the future. This activity will develop a wealth of process-specific knowledge and is likely to be a fairly well-rooted effort by the turn of the century.

In the following decade, we will almost certainly see the application of knowledge-based systems that employ this newfound wealth of knowledge to control existing processes, diagnose likely root causes of quality problems, and assist in new process design. These systems will also be a tremendous boon to training departments. This "intelligence" will be handled as a material by production and concurrent engineering professionals, who will be able to make it portable and transport it to other sites and other companies (such as suppliers) to improve their process quality. Eventually, "process intelligence units" may be called out as just another material listed in the bill of resources when a new product or process is installed.

- *Critical characteristics data.* Any given product, even an individual component part, has an infinite number of dimensional and other physical properties or characteristics. Most of them either remain constant through the production process or are not critical to the form, fit, or function of the product.

Some characteristics are vital, however, and must be closely controlled in each process. Identifying and monitoring these key characteristics (e.g., physical dimensions, hardness, acidity, color) will be the purview of shop floor process management. It will be these characteristics that define the "quality" of the product at each stage of the production flow. In most cases, these characteristics will be identified in the "flow-down" of customer requirements during quality function deployment as part of the concurrent engineering process. In other cases, the characteristic criticality will only become evident as its importance is recognized for downstream production processes.

Sophisticated Software on the Factory Floor

In future manufacturing environments, a far greater percentage of the "materials" used by production will be software. Various software is used on the shop floor today, including spreadsheet, data base management, and word processing packages. In addition, specialized software is used to run actual production equipment via programmable logic controllers (PLCs). In some of our more advanced facilities, small expert systems are used to perform such functions as preventive maintenance, scheduling, equipment maintenance diagnostics, and part "nesting" work. Simulation software is also becoming more widely accepted and applied for factory material flow and work load planning.

Future manufacturing operations will utilize much more sophisticated software in a plethora of applications. Some of the most important developments will be in the following areas:

- *Simulation software.* Future simulation software will be much more sophisticated, powerful, and user friendly. Such software used on the production floor will be able to move up and down through levels of abstraction, simulating operations at every level from multiplant to multimachine. The primary focus will likely be process and interprocess simulation to identify constraints and optimize and synchronize product flow.

• *Decision support systems.* Expert systems will grow in number and capability throughout the next two decades in order to support increasingly complex decision making. The balance of this decade will likely be spent in incremental module development, that is, the development of expert systems tailored around a specific need such as lot sizing.

As we enter the twenty-first century, these modules will be woven together in expansive decision support structures such as "intelligent" inventory management systems. Toward the end of that decade, knowledge-based system development work should provide the capability for these systems to "learn" and thereby apply fundamental "reasoning" logic to increasingly complex problems while considering more and more factors in their analyses.

Other developments associated with concurrent engineering and general changes in the way shop floor operations are managed will converge with these developments, so that the term *inventory management system* will probably become obsolete. It will certainly become irrelevant except possibly as a description of one or more subsystem reasoning modules. Inventory will be merely one of the many factors weighed in knowledge-based system analyses directed toward process and product line profitability. (Inventory management is used as an example here, but merely reflects the evolutionary changes that will occur in many of our existing software systems.)

• *Constraint management systems.* These will be knowledge-based systems that identify and record process interruptions and overall production disruptions, with their corresponding root causes. These systems will eventually "learn" from the data they collect and feed both the simulation software mentioned in the first paragraph of this list and the actual machine control systems embedded in the controllers of production cells and flexible machining systems. Related subsystems will track maintenance problems and manage preventive maintenance activity.

• *Process management systems.* These knowledge-based systems will initially track process capability, material properties, and the relationships that exist between them. They will create data bases from these relationships and identify correlations. Eventually, again probably around 2005, these systems should begin to "learn" from these relationships by inferring cause-and-effect, developing hypotheses, and testing them over subsequent production lots. Embedded in these systems will be known principles of materials science and mechanics that permit more advanced "reasoning" in the construction of hypotheses. As hypotheses are proved, actual knowledge may well be added to these sciences.

• *Materials development systems.* During the first decade of the twenty-first century (probably early in that decade), we should be using knowledge-based systems that support material design activities. They will support the decision making involved in this process by analyzing the physical properties required against known elemental structures and identifying desirable chemical configurations to satisfy those needs. Later, as concurrent engineering principles are applied more vigorously, the systems will learn to develop conceptual process requirements for fabrication of the materials, and eventually to compare those requirements to existing process availability, capability, and capacity. At some point, they will compare the costs of such configured materials and processes to the material value as a component of product cost and determine the economic feasibility of using the newly defined material.

• *"Teaching factory" systems.* These systems will "learn" from the data collected in the system modules described in the first and fourth paragraphs of this list. They will construct lesson plans and training outlines incorporating the original principles from the systems' initial knowledge base and the knowledge gained from actual production operations. The factory will be transformed into a learning environment where production operations and methods, equipment, and materials research are one and the same. Since the development of such systems is predicated on the system developments depicted in the first and fourth paragraphs of this list, these systems will probably not be available until between 2010 and 2020.

Tailoring Materials to Tomorrow's Manufacturing Needs

The third kind of material utilized by production is the physical material converted into actual products. The discussion in Chapter 6 on the materials currently used in predominant manufacturing operations included the following:

- Metals and alloys
- Plastics
- Ceramics
- Composites
- Paper
- Fiber
- Wood
- Stone

These broad generic categories are still serviceable in terms of encompassing the spectrum of raw materials commonly used. However, that is about the only thing that remains largely unchanged in the materials arena.

As we move toward the close of the 1990s, we find ourselves at an important decision-making juncture pertaining to new materials and the competitiveness of American manufacturing. Only if materials scientists and engineers integrate with the rest of business operations will U.S. firms improve their competitive positions in domestic and international markets. In *Materials Science and Engineering for the 1990s,* the National Research Council says:

> Materials science and engineering is crucial to the success of industries that are important to the strength of the U.S. economy and U.S. defense.
>
> Greater emphasis on integration of materials science and engineering with the rest of their business operations is necessary if U.S. firms are to improve their competitive positions in domestic and international competition.
>
> Scientists and engineers must work together more closely in the concurrent development of total materials systems if industries depending on materials are to remain competitive.

These factors point to a trend that is clearly going to distinguish winners and losers in future manufacturing. Following is an examination of manufacturing materials categorized by most obvious future application.

Structural Materials: From Metal Matrix Composites to Microcracked Ceramics

The most important properties of structural materials include strength, toughness, weight, hardness, and stiffness. The weight-to-strength ratio is likely to remain a central theme in measuring our progress in developing and utilizing structural materials. Our current technology for using existing materials, especially the more traditional metals, is rapidly approaching its limits in this area.

Therefore, we will be forced to explore more exotic materials and improve our material synthesis and production technologies dramatically to make significant product breakthroughs.

Metals will continue to account for a major portion of the structural materials used through the end of the 1990s. However, the kinds of metals utilized will expand substantially and will include the following:

- *Intermetallic compounds,* which offer high thermal resistance and very good load-bearing properties
- *Fine-grained, single-phase metals,* which offer high levels of resistance to corrosion, good ductility, and excellent strength; will become more commonly used as we improve rapid solidification processes
- *Metal matrix composites,* which offer very high weight-to-strength ratios and enormous shape flexibility
- *Microduplex alloys and dispersion-strengthened alloys,* which offer high levels of thermal resistance and great strength in multiple axes
- *Radiation-resistant alloys,* whose primary benefit is obvious from their name

Ceramics will grow significantly as a percentage of the materials used in manufacturing over the next two decades. Ceramics offer low weight, high levels of compressive strength, extremely high levels of thermal resistance, good corrosion resistance, and several other important properties. In fact, were it not for the problems of brittleness and manufacturing cost, ceramics would appear to be a nearly perfect structural material. Although the base material itself is nearly always the same, it is modified in a number of ways in order to deal with the problems of brittleness and cracking:

- *Transformation-toughened ceramics* improve the toughness and crack resistance of the material by altering its crystalline structure.
- *Fiber-reinforced ceramics* accomplish this same purpose through embedded fibers within the material.
- *Microcracked ceramics* achieve crack resistance by incorporating minute cracks in the surface of the material (precracking).

It will also be increasingly common to encounter ceramic matrix composites that, like the metal matrix composites described earlier, offer increased strength to the composite structure with minimal weight and increased form flexibility.

Polymers may well have the greatest potential for short-term growth in terms of manufactured product share. They offer some extremely useful characteristics for our future manufacturing needs, including high strength-to-weight ratios, outstanding formability, and good thermal resistance, electrical resistance, and optical properties.

Of the existing polymer market, which had already surpassed $100 billion by 1989, structural polymers are a major component. These applications generally involve aromatic polyamide polymers or polymer-fiber composites as structural component materials, and polymer adhesives. Structural polymer-fiber components commonly employ carbon fibers, ceramic fibers/whiskers, kevlar, or glass fibers/whiskers immersed in resin. The parts are generally cured in ovens or autoclaves to harden them and then may be machined like their metal predecessors.

Composites are particularly interesting substances because of their superior weight and strength characteristics. However, the actual fabrication processes involved with these materials are often still more expensive than those of similar metal components. Therefore, additional work will be required (probably several more years' worth) before they gain the position of prominence for which they are destined.

Electronic Materials: Polymers, Silicon, and SQUID

Electronics materials generally include polymers, semiconductors, magnetic materials, insulators, and metals. This field is likely to remain among the most aggressive in terms of materials development through the end of the twentieth century because of the incredible competition for speed in information processing.

The primary element used today, and likely through the balance of the 1990s, is the silicon semiconductor. Silicon in this application offers the important properties of availability (great natural abundance equates to relatively low cost) and high degrees of crystalline alignment. The alloy silicon-germanium is also in limited use in electronic materials, as is gallium arsenide. Other elemental materials currently in development that have met with less widespread success in this area include combinations of indium and/or aluminum with elements such as arsenic and phosphorus.

Some of these will evolve into useful replacements for silicon in specific kinds of applications, but the primary thrust in this area for the remainder of the current decade will likely continue to center on the development of progressively smaller electrically isolated areas of single-crystal silicon.

Ceramics are often used as substrate material devices requiring high levels of chip density and circuit density, or where mechanical strength and structural stability are important. Alumina is the most commonly used ceramic substrate material today, and is generally utilized in combination with molybdenum conductor materials. It is likely that new substrate/conductor combinations will be developed by the end of the 1990s, because of inherent difficulties associated with poor transmission characteristics and other problems arising from high-temperature fabrication methods.

Recent substrate material developments have focused predominantly on polymers. Polymer materials allow fabrication at lower temperatures, can be manipulated to control surface flatness to a higher degree, and support the use of lithographic processing in the fabrication of metal lines for electronic signal transmission. There are some difficulties associated with polymer substrates as well, most of which involve heat dissipation and low levels of mechanical strength and structural stability. New polymers and ceramic/polymer combinations that incorporate lower thermal expansion coefficients, higher cracking resistance, moisture resistance, and photosensitivity will probably become available around the end of the 1990s.

The polymeric materials most commonly utilized as the chip-carrier medium in current electronic devices are constructed of epoxy-impregnated glass cloth. The glass fibers provide strength, while the polymer provides a structural housing. This material will likely be enhanced by replacing the glass fibers with aromatic polyamids. Other enhancements will also be made in conjunction with the polymer developments outlined in the previous paragraph. These improved combinations increase structural integrity and the capacity to handle higher levels of wiring density.

Other electronics-related applications of ceramics and polymers that have been identified as candidates for development over the next two decades include these:

- Capacitor coatings for substrates and chips
- Photosensitive polymers for photolithography
- High-strength polymers spun from liquid crystal solutions
- Piezoelectrics, especially piezoelectric polymers

Metals traditionally utilized in electronics applications have been widespread, including everything from sheet metal housing materials to electronic signal transmission lines. Today, the most important applications are transmission lines and magnetic film used to store information. Metal wiring is still used to make the electronic connections between the individual layers of each computer chip, and between the chips and other electronic devices as well.

Because of the increasing density and decreasing size of these components, the sizes of wiring are being reduced to a level where reliability is a serious concern. Current efforts in terms of research in this area include reliability of the connections between wiring and semiconductors, wiring and metal layers, wiring and contacts, and wiring and chip-mounted or substrate-mounted devices. Other developments being pursued and likely to be realized to some degree before the turn of the century include wiring that is more reliable at smaller sizes, wiring supporting finer levels of photolithography, and wiring materials that are more compatible with insulating materials.

Magnetic materials are an extremely important element of the electronic materials field, and are absolutely critical for information storage. The materials used to date have been almost entirely composed of iron and chromium oxides in storage media and permalloy or ferrite heads. As we move into the second half of the 1990s, a new class of magnetic materials will likely be exploited. These materials are known as spin glasses, and the statistical theory of magnetic behavior that drives the interactions of these materials is far different from that for traditional magnetic materials. In fact, the concepts of complex magnetic behavior employed in spin glasses led to the concept of neural nets.

Superconductors show an extremely steep growth curve as we enter the second half of the 1990s. In fact, this segment of the electronics field, which involves complete systems such as superconducting quantum interference devices, or SQUID, now does around $1 billion in sales. Currently employed superconductor materials include niobium titanium and niobium tin compounds. Recently, a group of ceramic oxides containing lanthanum, barium, and copper were used to improve superconductivity temperature levels, as was a compound of yttrium, barium, copper, and oxygen known as "123" and other compounds of thallium, barium, calcium, copper, and oxygen. These and other compounds like them will continue to be developed in laboratories and put to use over the next twenty years.

Photonic Materials: Just Add Light and Process

Photonic (or optic) materials will be utilized in many of the same ways that electronic materials are used today. They will be used in conjunction with light (rather than electricity) to transmit, convert, store, and process information. Indeed, in the form of fiber optics, these activities are already well under way. Beyond fiber optics, semiconductor lasers that operate at room temperature, low-noise detectors, and optical and optoelectronic circuits are also already in use. These materials are employed primarily in electronics applications. However, photonic materials such as amorphous silicon are also used in devices such as solar cells. Other photonic materials are employed to produce devices such as ultraviolet lasers for laser fusion operations. Still other materials, such as erasable optical memory material, gradient-index coatings, flat-panel displays, and optical window materials, will have significant impacts on several manufacturing industries as we enter the next decade.

These and other optical materials are among the most likely near-term avenues to achieve microprocessing and microfabrication capabilities. The challenge in this area over the next couple of decades is vapor-deposition control in multiple dimensions for materials fabrication.

Finally, in terms of photonics materials research, we are likely to see significant efforts in an area dubbed "microprogrammable materials." In *Materials Science and Engineering for the 1990s* (National Academy Press, 1989), the National Research Council Committee on Materials Science and Engineering says:

> Serious consideration should be given to a research program aimed at what might be called microprogrammable materials. The phenomena of spectral hole burning and photorefractive effects represent limited examples of this idea. It might be possible to use energetic beams to make localized changes in the chemical bonds or the structure of a material in a way that will change its function at desired points in space. For example, it might be possible to convert an insulator into a semiconductor or a semiconductor into a superconductor in spatially controlled regions.

Biomaterials: The Stuff of Life

Biomaterials are commonly viewed as those materials used to construct artificial organs, biochemical sensors, drug delivery systems, dental devices, and so on. Some of these materials have been successfully implanted in living humans, which has spurred demand in a few specific material categories. However, manufacturing costs and application procedures are still very expensive, and are thus inhibiting the growth of the use of these materials.

New synthetic biomaterials are an area of considerable interest, and the primary challenge (aside from cost) in this work centers on improving the compatibility of the new materials and the biological tissues with which they interface. Corrosive properties, degradability, and durability are also significant factors that must be considered as these new materials are developed. New synthetic materials under development include polymers and ceramics. Other naturally occurring and heavily modified elements include both metals and carbons.

Beyond the synthetic materials utilized, biologically derived materials are also already in fairly widespread use. These materials include processed tissue from both pigs and cows, human umbilical veins, and reconstituted collagen and elastin. Among the research areas that will have a significant impact on the medical device manufacturing community are tissue adhesion control, protein and cell removal/manipulation, and immunocompatible surface property enhancement. One of the areas under development today is polymers with affinities for specific proteins. Over the next two decades, the combination of synthetic and biologically derived substances will likely be the primary thrust in the biomaterials field.

The Materials Used in Distribution

The materials used in distribution in the next century will expand from the data and packaging materials described in Chapter 6 to include software as well.

The materials used in distribution.

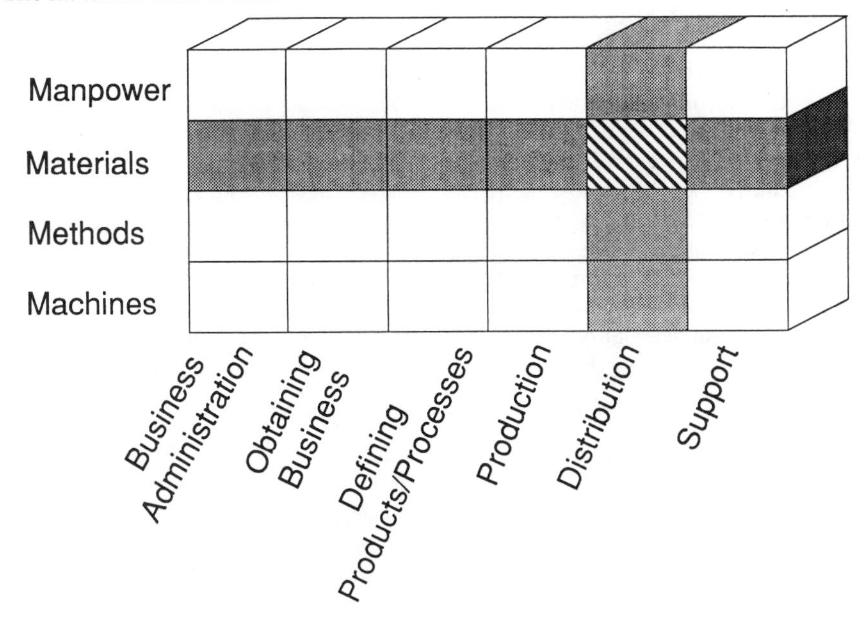

Getting Off the Paper Trail

Data will be grouped into categories similar to the ones used today, including the following:

- Delivery schedules
- Route data
- Third-party data
- Inventory data
- Bar code/ID data
- Market/sales forecasts
- Cost data
- Field performance data
- Delivery performance data
- Packaging material data

The data itself will change somewhat. The media through which it is presented will evolve from the normal paper base (e.g., paper maps, invoices, and packing slips) toward an electronic base.

Maps will be displayed on CRTs and flat panel displays, even onboard transport vehicles. This will provide a means to continuously update the information used, accounting for traffic delays, weather patterns, and road conditions. Applying satellite, cellular telephone, and other new communications technologies, these devices will also perform communications functions. They will provide constant vehicle/load location visibility and a means to redirect some deliveries en route.

One of the real plagues of receiving departments today is material that arrives without the appropriate accompanying documentation. With the advent of applied high-tech electronics, this situation will disappear. Fax equipment has already been made portable and usable in conjunction with cellular phone equipment for written verifications and approvals at any site. Inspection of paperwork can occur via fax or video display of EDI data from the supplier before the shipment ever leaves the supplier's facility.

The increased levels of strategic alliances, mergers and acquisitions, and other similar consolidations will likely mean that combined data such as delivery schedules will also become an important element of the information utilized in distribution of finished goods. This will make shipment configuration and route planning a more challenging activity. When multiple manufacturer distribution channels are undertaken in conjunction with multiple delivery points and perishable or limited shelf life products, the complexity will be even higher. At this point, expert systems are likely to come into play.

For those manufacturers electing to procure their distribution functions as outside services, these complex problems will be avoided. However, they will still have significant impacts on the cost of the services, the quality of the delivered product, and the overall financial health of the company. Other manufacturers will elect to retain control of these processes, and those companies will resolve many of the resulting challenges.

The software used to manage distribution operations over the next decade will likely be developed in several stages. The initial stages will involve selecting the best transportation mode for individual shipment quantities and costs. Additional stages will likely focus on transport vehicle routing, delivery scheduling, and load-sizing. Around this same time, concurrent engineering efforts related to packaging strategies will provide enough information to develop knowledge-based systems for the design of specialized packaging and for optimum application of standard packaging materials.

Eventually, probably in the first half of the next decade, a complete knowledge-based system will be developed to provide decision support in all of these distribution areas, integrating design and deployment of packaging, sizing and configuration of loads, selection and routing of transport methods, optimization of distribution center and factory-level finished goods inventories, and load identification and tracking.

Finally, a few years later, the knowledge-based system will include performance data collection capability and be integrated to other factory systems, so that accounts receivable transactions, sales forecasts, and so forth can be updated in "real time" from the points of delivery.

Predictions on Packaging

Packaging and shipping materials are beginning to incorporate new materials such as composites and other polymers. They are also being designed to withstand the hazardous properties of modern production materials and their by-products. Some developments in the materials we are likely to continue to utilize through the balance of the 1990s are the following:

- *Containers.* It seems safe to say that as we continue to apply advanced materials to the containerization of our products, container wall thicknesses will diminish, container weight will diminish, cost per unit shipped will diminish, and protection levels will improve. Biodegradability and recyclability will also play an increasingly important role in container design and selection. It also seems reasonable to imagine that as technology allows increasingly small computer chips that are more effective and less expensive, future containers will have embedded programmable chips to speed identification of the load, its point of origin, its destination, and its contents.

- *Cushioning materials.* Perhaps the most interesting innovations in cushioning materials being developed today are those that utilize the least expensive and most flexible element—air. It would be difficult to imagine a more simple and inexpensive medium for cushioning and insulating materials than packing materials made up primarily of air bubbles.

Another fascinating application is the flowable medium for air bubble cushioning. In one form, this material is shot into the container under and around the protected item in a viscous liquid form, then "rises" in a manner similar to the way a foam "head" rises at the top of a newly poured carbonated drink. The foam then solidifies in its risen condition, filling the voids between the product and its package with millions of tiny pockets of air.

Other forms of cushioning materials rely on fibers (sometimes referred to as "rubberized hair") that may be applied in a free condition or molded to fit the product dimensions, pellets (sometimes called "peanuts"), styrofoam, and convoluted preformed polyurethane foam.

Over the next several years, additional chemical and physical property enhancements will be applied to these materials, incorporating characteristics such as fire retardation, thermal resistance, structural integrity improvements, and selective biodegradability.

• *Barrier materials.* Barrier materials are used in current packaging operations primarily to isolate products from humidity and other moisture. The most widely employed materials include polyethylene bags, film, tubing and tarps, and volatile corrosion inhibitor (VCI) in paper and bags. It is likely that these materials will continue to be used. Many new materials (such as "pre-preg" composites) require tight control of their chemical properties as well as thermal control when transported in their pre-cured state.

• *Shielding materials.* Magnetic shielding materials are currently used to protect products and product components from electrostatic discharge (ESD) damage. Although initial applications were focused almost entirely on the protection of metal oxide semiconductors, potential applications continue to grow.

As our products grow in complexity and "knowledge" content, we will likely see at least moderate growth in the use of these materials. Eventually, our products themselves are likely to become less susceptible to such environmental hazards and require no such specialized protection. However, it seems reasonable to estimate that this won't occur for at least several years. This kind of packaging material is currently applied to the packaging of microwave devices, circuit boards, computer memory units, thin film resistors, and so on. Typical shielding materials currently utilized include antistatic polyethylene sheets, film and bags, antistatic polyethylene foam, antistatic polyethylene bubble cushioning film, conductive shuntfoam and polybags, and polypropylene jars. In products requiring protection from electromagnetic interference (EMI), radio frequency interference (RFI), standard magnetic interference, or electrostatic discharge, metal foil is sometimes utilized also. These materials, like the other shielding materials described, are likely to find additional applications through the balance of the 1990s as our products become more electronically dense; they will later diminish as shielding materials are applied within the products themselves and additional packaging protection becomes superfluous.

The Materials Used in Product and Customer Support

Over the next two decades, software and data will be critical success factors in product and customer support for American manufacturers. Existing data will change in substance, and new kinds of data will be required. Software will be developed, modified, and enhanced, and there will be a move from passive software to knowledge-based systems. These systems will initially provide decision support information, and eventually generate products of their own. For example, imagine software that collects data regarding field failures as they occur, spots causal relationships, analyzes frequency, and develops draft bulletins to be issued to field repair and sales/service organizations. All the human supervisor is required to do is verify the wording and approve the service bulletin. The system then generates a targeted mailing, or utilizes E-mail to distribute the document.

A great deal of data is already employed in product and customer support, including service part sales data, repair response time, repair parts inventory, repair equipment records, customer service crew management information, service

The materials used in product and customer support.

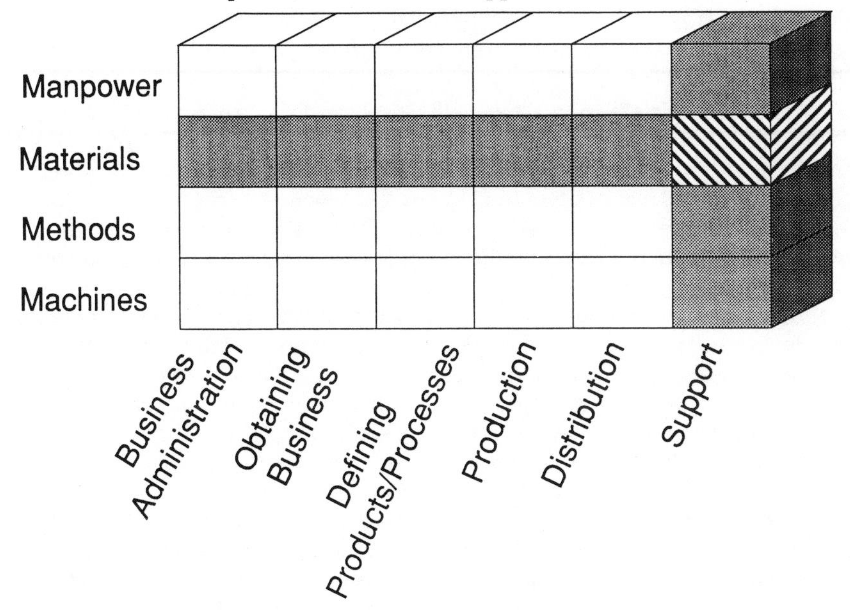

agreements and warranties, technical publications, repair manuals, and training materials. In the future, additional information will be utilized alone and in conjunction with existing data to support customers and products in the field, including the following:

- *User survey data.* In the spirit of quality function deployment (QFD), users will be surveyed frequently to determine the aspects of customer and product service that are important and to measure the degree to which these requirements are satisfied by existing programs. These surveys have been utilized sporadically at best by product/customer service organizations through the first part of the 1990s, but are likely to become an increasingly prominent aspect of future service programs.
- *Benchmarking data.* The best customer and product service organizations already employ benchmarking of their services against those of direct competitors. However, relatively few benchmark themselves against others who, although not competitors (in fact, in most cases aren't in the same business at all), are the "best in class" at supporting customers or products. In the last half of the 1990s, benchmarking will become a far more pervasive activity, and by the turn of the century it may well become a common practice.
- *Customer advisory team feedback.* Using teams of existing and potential customers has long been a tool of good marketing programs. However, it has been less widely applied in other areas of the business, including customer and product support. Such teams can provide invaluable information regarding how a service organization should be trained, equipped, deployed, and measured. They can also

act as a sounding board for new service and support ideas, evaluating them before much money is spent in deploying ideas that may not be accepted.

As we move through the 1990s, we are likely to see use of these teams increase dramatically.

• *New design data.* Because there will be substantial customer/product support involvement in the concurrent engineering process, this group will have access to new design data much earlier than has traditionally been the case. As a result, they will be in an ideal position to field-test the new design concepts on a what-if basis with customers, offer special prototype features to customers who already own existing products, and report their findings back to the concurrent engineering teams while new products are still in development and initial deployment stages. In addition, customer/product support professionals are often the first to identify new customer needs in terms of product features, maintainability, packaging, and usability. This link to concurrent engineering efforts will be one of the most potent factors in development of new products, reductions in cycle-to-market times, and overall market acceptance (and therefore market share).

As we move through the end of the 1990s, this information will continue to be gathered more formally through survey forms, structured interviews, and so on. Toward the end of the decade, many companies will maintain electronic files of the data that engineers and concurrent engineering teams may refer to as they see fit. As we move into the first decade of the next century, these companies will turn into a science the collection, analysis, and incorporation of this information into new designs. It will simply become the way business is done. All other factors being equal, those companies who develop the most effective and efficient means of performing this activity will almost certainly own the dominant shares of their markets by 2010.

• *Service center management data.* Service centers perform a major portion of all service on many manufactured products, for example, automobiles, aircraft, and appliances. Because of this activity, these facilities are uniquely capable of gathering and reporting trends in service requirements and capability that may remain transparent to both the factory and individual customer/product support personnel. Trends in such areas as maintenance frequency, product failure causes, and failure severity are often most visible and recognized earliest in product service centers.

However, as with new design data, today's service centers are generally not well equipped or structured to aggregate the data, turn it quickly into useful information, and effectively communicate that information to the affected concurrent engineering teams. As we move through the end of the 1990s, this will change as data bases are built, then linked electronically to concurrent engineering organizations. Around the turn of the century, knowledge-based systems will be developed that apply artificial intelligence to the aggregation and analysis processes. These events will increase the responsiveness of manufacturers to field problems in several areas, including distribution of service parts, deployment of service personnel, and development and deployment of service training.

• *Customer training data.* The training available to customers in terms of how to use their new products, and new ways to use their old products, is an invaluable source of potential revenue and market share retention. Especially as the customer/

product sales records from point of sale are matched against current training records, customer/product support personnel will be able to mount extremely effective targeted training efforts. Because the training will be tailored to provide support that is truly needed and targeted to exactly those customers who can benefit from it, this kind of information will be a real revenue generator to the companies who use it well.

This data is generally among the least organized and most poorly utilized in many American manufacturing companies, partly because of third-party training, but mostly owing to a simple lack of appreciation for its importance and revenue-generating potential. As the use of relational data bases, electronic data interchange (EDI), and then knowledge-based systems becomes increasingly prevalent, this situation will change. It is likely to follow the same developmental pattern as the new product design data and service center management data, in that data base development and communication link development will precede effective use of the information, and should occur in mainstream companies by the end of this decade. Application of artificial intelligence to this process is probably about a decade away.

- *Product performance data.* Product performance data is among the most widely gathered, scrutinized, and published data used by the customer/product support organization, as well as by competitors, customers, potential customers, trade publications, and advertisers. (Publications like *Consumer Reports* have become a mainstay of the American consumer and are widely recognized to provide an extremely valuable service. They provide effective, objective, and comprehensive comparison data about competing product offerings for potential customers.)

Several advances regarding this data are likely to occur. It will become more reliable, will be more readily available earlier in the product life cycle, and will be utilized more effectively than it has been to date. It will become more reliable as the devices for maintaining this data become increasingly sophisticated and more frequently embedded in the products themselves. It will be more readily available via EDI and computer-based gathering and aggregation equipment/systems. It will be correlated to simulation-based fatigue and other performance models so that warning signs are more quickly recognized. It will be more effectively utilized as these factors yield more objective, real-time information and present it in usable forms to concurrent engineering and customer/product support professionals.

- *Technical support/service data.* The data fed back from field technicians and other field service professionals is also usually quite spotty today. Field service people are generally technically skilled, but often lack the training, skills, equipment, and direction to record and report this information in a cogent, organized fashion. Service information is generally recorded in a brief and cryptic manner on a one-page service record. The service record is generally put into a drawer somewhere and referred to infrequently thereafter, e.g., to audit service revenue records.

As service equipment continues to become more sophisticated and information-based, it will also incorporate more capability for information gathering and recording. It will therefore be almost effortless to periodically review service equipment records, determining the types of repair and maintenance work being performed, in what locations, and in what volumes. This information will then be

Exhibit 13-2. Future customer/product support system structure.

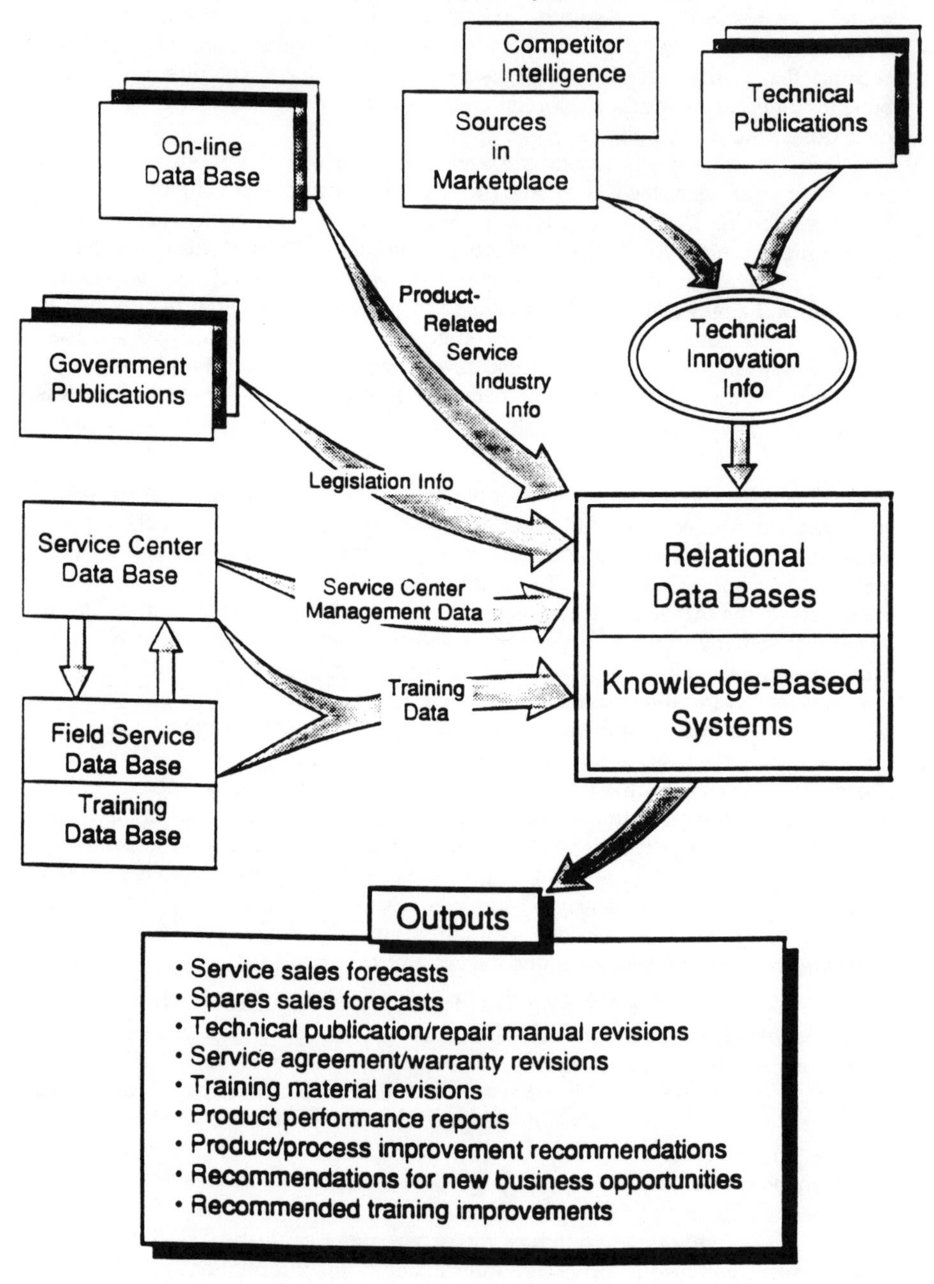

fed via EDI to service centers or directly to the manufacturing site for aggregation and analysis.

• *Infrastructure data.* Most manufacturing companies still do not understand what infrastructure information is. Among those who do, the majority have not recognized the need to gather and use such information. Most of those who have recognized the need have not yet developed the systems and processes to perform this activity. It will probably be sometime after the turn of the century when this concept becomes a real working process in most mainstream American manufacturing organizations. Infrastructure data such as product usage platform data, relevant legislation information, product-related service industry data, related technical innovation information, and user group feedback will need to be gathered and properly aggregated/analyzed to be used effectively in customer/product support as well as concurrent engineering. This is another area that requires integrated development of several processes and systems.

We can see that there is a huge opportunity for both hardware and software developers in the area of customer/product support over the next two decades. As data gathering and management becomes the primary enabler of leading-edge service, it will also become a primary discriminator between the winners and the losers in this competitive aspect of revenue generation, customer satisfaction, and, ultimately, market share. The information, hardware, and software will make possible systems that reflect the relationships depicted in Exhibit 13-2.

Summary

Materials science has been described as one of the three "megatechnologies" most likely to transform the world as we know it over the years ahead. Both products and processes will be profoundly changed by the sweeping effects of this transformation. Imagine automobiles that are just as strong and rigid as today's models, but are made from material the thickness and weight of standard household aluminum foil, and you get a pretty clear idea of where we are headed in the area of physical material. Beyond this, when manufacturing people think of "material" in the not-so-distant future, many of them will think as often of data as physical substances. Ultimately, the lines between materials, intelligence, and even packaging in some cases will become so blurred as to make it virtually impossible to separate one from another.

Chapter 14

Dominant Machines: Equipment With a High IQ

When we considered new equipment a decade or so ago, the leading edge was numerically controlled machinery. The acceleration of intelligence applications in manufacturing equipment since then has been incredible. Consider this excerpt from an advertisement for a device that may be attached to operate complete machines, which comes from a recent issue of a popular manufacturing equipment catalog:

> . . . combines built-in logic, timers, counters, and built-in operator interface—all in one box. The easy-to-use program comes with 40 I/O, 80 timers, 46 counters, and 2K ladder code. It also offers multi-drop RS-485 port, motion relay, resolver position input, six programmable limit switch relays for high speed precision control, 16 control inputs, and 12 programmable logic control relay outputs. Multiple program selection permits quick product changeover. The Model M1500 can be programmed from an IBM-compatible computer.

As we move through the end of the 1990s and into the next decade, more and more intelligence will be built into equipment, and the operation of equipment will continue to become more automated. In addition, it will become increasingly difficult to separate software from hardware in this environment.

From business administration through customer and product support, the computer will continue to be the focal point of dramatic changes in the machines used to perform and manage manufacturing operations. In this chapter, we examine some of the likely developments in these areas. (See Exhibit 14-1.)

The Machines Used in Business Administration

One of the most important challenges for business administration professionals in the 1990s and in the first decade of the next century will be the integration of all of the computing and telecommunication technologies of their business into a single integrated information system.

As we approach the middle of the 1990s, senior executives across all manufacturing (and nonmanufacturing) companies have begun to ask repeatedly and pointedly about the ROI of their information technology expenditures. Operating budgets for these systems have reportedly grown at an average of about 10 percent per year over the last several years, in an environment where "downsizing" and "streamlining" have simultaneously become the dominant operating philosophy.

Exhibit 14-1. Dominant U.S. manufacturing machines.

U.S. Manufacturing Machines					
Business Administration	**Obtaining Business**	**Define Products and Processes**	**Production**	**Distribution**	**Support**
• Integrated workstations • Database management systems • Servers • Application-specific departmental systems • Mainframe/super-mini/mini/personal/notebook computers • "Workplace in a briefcase" • Wide area networks • Local area networks	• Personal (notebook) computers • Modems • FAX machines • Metropolitan area networks (MANs)	• Mainframe-level workstations • Relational databases • Knowledge-based design systems	• Extremely sensitive sensors and monitors • Enhanced robotics applications • Intelligent control systems • FMS equipment • Nontraditional cutting/finishing equipment • Advanced coating equipment • Advanced fixturing • "Smart" robots • Imbedded diagnostic equipment • Natural language processors • 3-D model display equipment	• Automated material handling systems • Automatic storage/retrieval systems (AS/RS) • Vehicle-based navigation systems • Automated packaging equipment	• Telecommunications gear • Desktop publishing equipment • Video, audio equipment • Interactive multimedia equipment • "Knowledge-based" troubleshooting equipment • Remote diagnostic and repair equipment

The machines used in business administration.

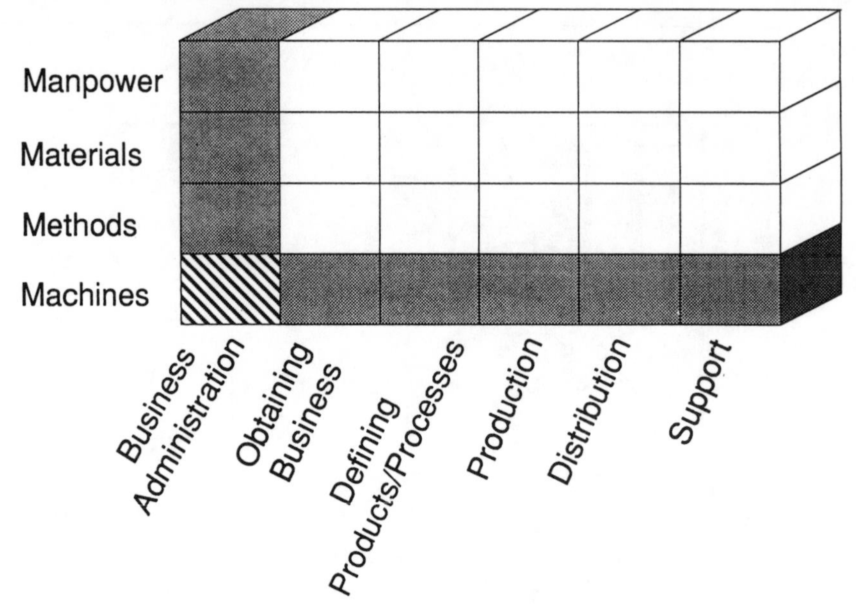

This combination of events has generated intense scrutiny of the information technology used by manufacturers and open skepticism on the part of many. The situation is most critical in those organizations where poor strategic information system planning has been done. Incompatible individual application systems (both internally developed and purchased from external sources) and growing support budgets for each of them have resulted in very low gains in productivity and very low levels of strategic advantage when compared to the money and time expended, forecast productivity, and anticipated competitive advantages.

Recent studies indicate an average job expectancy of about 2.5 years for chief information officers (CIOs), which is no doubt one result of this problem. However, the role of CIO may well remain, even if under a different title. These functions, at least, will need to be performed at that level:

- *Information network management* to facilitate creation and maintenance of consistent interfaces and connectivity standards
- *Maintenance and enhancement* of system structural elements
- Establishment, maintenance, and enforcement of *cross-functional systems standards*, such as EDI standards
- *Oversight of research and development* of new systems/applications for end-user organizations

Given the need for these functions, what kinds of computational equipment will need to be orchestrated in future manufacturing environments? The evolution of this equipment over the last several decades holds some important clues.

By 1960, virtually all computing systems were batch-processing systems.

Voice communication was handled by integrated long-distance networks. Facsimile systems had come into limited use. At the end of that decade, some on-line systems were in use, a few stand-alone word processors had sprung up, and communication satellites had been deployed.

During the 1970s, on-line systems became the rule rather than the exception in mainstream manufacturing environments. Personal computers entered the scene, word processors were beginning to be "clustered" together, and toward the end of that period, local area networks (LANs) were in use.

In the 1980s, we saw the advent of major distributed data processing structures, which (combined with the PCs, clustered word processing equipment, and emerging distributed data processing structures) led to general network processing. Voice and data PBXs completed the information/telecommunication system foundation built in that decade, leading to what industry experts believe will become the major thrust in this area during the next decade: integrated information systems.

The Advent of the "Seamless" System

The primary focus of information technology professionals over the next decade or two will be the development of an integrated information system for their enterprise that is characterized by complete integration of systems (seamless or transparent access from any system to any other system's data); complete data/information integrity (which necessarily means timeliness); maximum resource utilization (minimum operating and initial investment costs); robustness (ability to readily and successfully adapt to changing user requirements); disaster resistance and recovery capability; and system data security. (To envision the information and telecommunications system architecture of the next decade in mainstream manufacturing organizations, consider Exhibit 14-2.)

Most mainstream manufacturing companies throughout the 1990s will utilize this kind of computer system structure in their business administration operations:

- *Integrated workstations* utilizing MS/DOS, OS/2, or Apple operating systems
- *Data base management systems,* typically Oracle or DB/2
- *Servers,* with Unix or OS/2 EE operating systems
- Some *application-specific departmental systems,* including one or more of the operating systems in this list, or another system such as the AS400

A great deal of insight in this area can be found in A. D. Little's *Forecast on Information Technology and Productivity,* which predicts the following:

- IBM's System Application Architecture (SAA) will be adopted by most users and third-party software vendors.
- MVS/ESA will virtually always be the operating system employed in mainframe environments.
- The most popular data base manager will be DB/2.
- The most likely basis of departmental systems will be application-specific superminicomputers. IBM's AS/400 product line is likely to be the most widely utilized.

Exhibit 14-2. Comparison of typical integrated information systems architecture with typical communications system architecture.

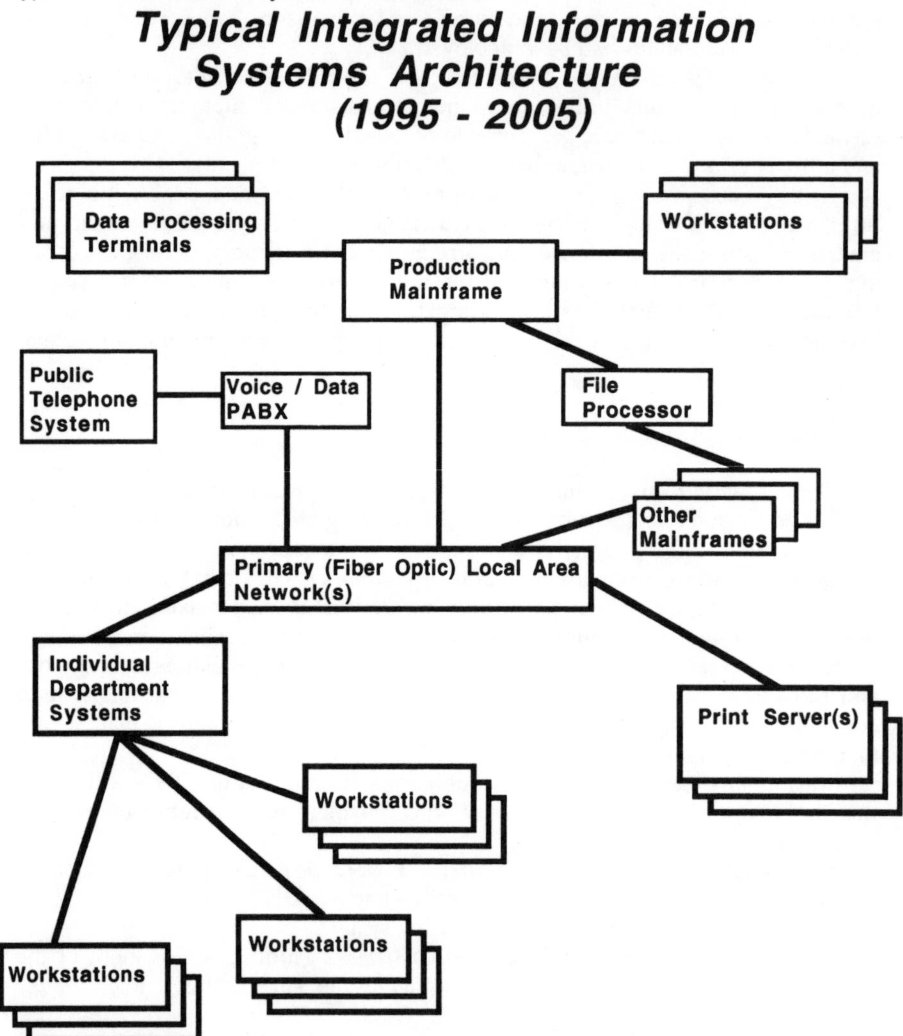

Typical Communication System Architecture (1995 - 2005)

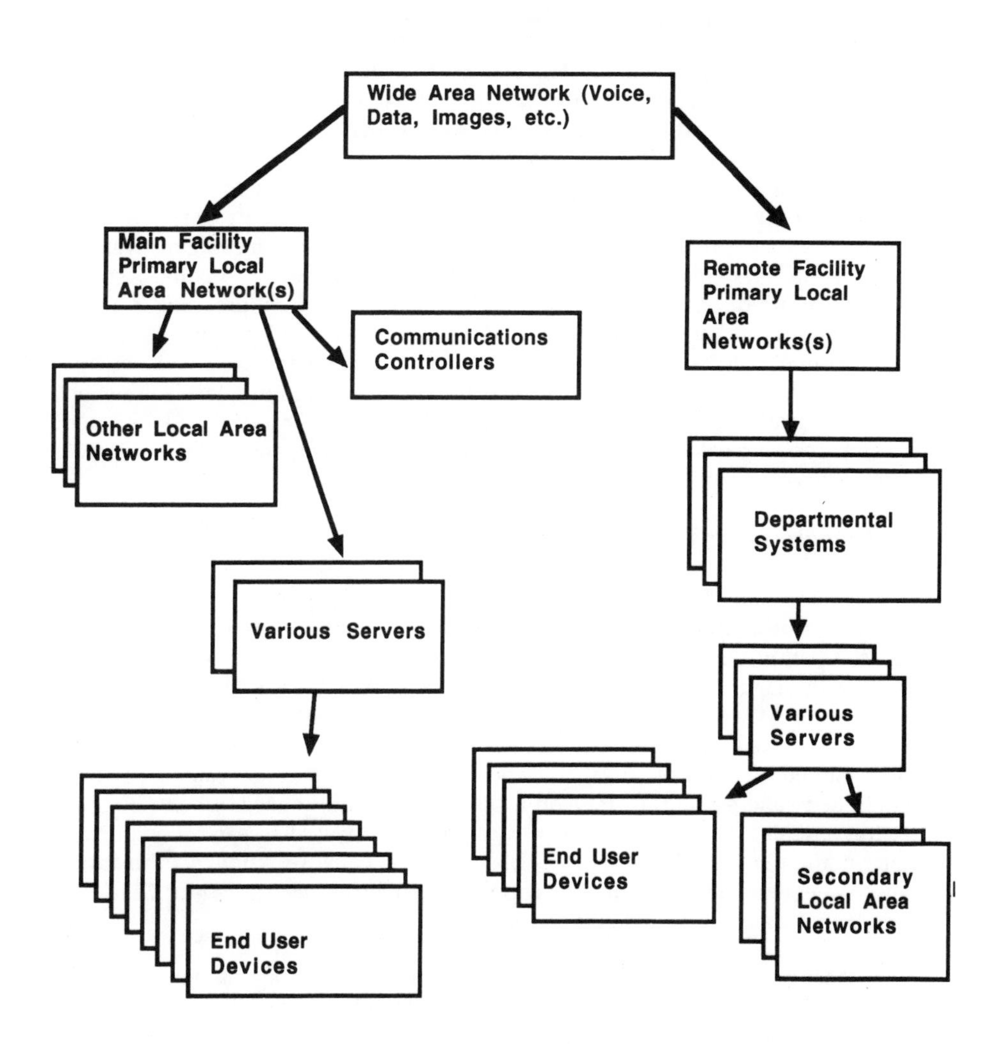

- New multimedia applications based on OS/2 will emerge. These applications will combine sound, video, graphics, and data. (Likely first applications include training, sales presentations, and executive information systems.)

In terms of hardware cost versus performance, it has been estimated that the price of high-end mainframes, which are likely to remain popular with mainstream manufacturers, will drop from their 1990 levels of about $100,000 per MIP in 1990 to about $20,000 per MIP by 1995. Smaller units will provide even more improvement in their cost/performance ratios.

Another factor to consider in terms of computer hardware and telecommunications equipment is the expected growth in telecommuting. With increasing numbers of employees performing work in their homes, and with executives more mobile than ever before, a "workplace in a briefcase" approach will almost certainly become more popular. Such a system is likely to include a personal computer, EDI functionality in both telephone and facsimile transmission and receipt, hard-copy printing capability, and rechargeable extended-use batteries.

Office automation is likely to continue to accelerate over the next ten years, with mainframe computers handling the majority of document storage and retrieval functions. They will also likely act as the general "traffic controllers," through communications control devices, of electronic mail and other messaging systems. Departmental systems will handle, as many do today, work schedules, special (manufacturing-day-based or accounting-day-based) calendars, program and project scheduling, and other miscellaneous departmental system functions. Generally, departmental systems will be the repository of local-use data bases, while companywide data bases will be handled at the company mainframe level. Functions such as general office automation, artificial intelligence applications, and document image processing will be handled by the mainframe as well, and will be integrated as they are developed or procured.

Biocomputers: From Atoms to Apples

At some point (probably beyond the next twenty years), we will be dealing with the incorporation of biocomputers into our systems architecture. Biocomputers, computing devices with some of the features of living organisms, promise to provide staggering computational power and speed. At a recent lecture on "Computing Machines of the Future" in Japan, Richard Feynman stated that there are few physical limitations to the speed of computation, since a bit of information can be stored on one atom, heat efficiency related to computation can be increased indefinitely, and signal transmission is limited only to the speed of light. Building on this premise and other related theoretical and physical research, Tsuguchika Kaminuma and Gen Matsumoto published an exhaustive study on the future development and deployment of biocomputers. In *Biocomputers, The Next Generation From Japan* (Chapman and Hall, 1991), they state:

> Biocomputers will gradually emerge from the convergence of four main streams of research: computer science, life science, architecture/software research and developments in biodevices.

The computer science development path will move through large-scale integration, very large scale integration, molecular computers, and atomic computers. The architecture/software development path will move through parallel architectures, pattern recognition equipment, and neurocomputers. The biodevice development path will move through protein engineering, biosensors, and a broader spectrum of biodevices. The life science development path will need to move through genetic information analysis, a great deal of developmental research, and eventually through neuroscience research before it can lead to biocomputer production.

These developments will correspond closely with the emergence of microfabrication, nanofabrication, and nanoassembly discussed in Chapter 12.

The Machines Used in Obtaining Business

Like their business administration counterparts, the machines utilized by manufacturers in the area of obtaining business over the next decade will consist primarily of computer and telecommunications equipment. When the types of equipment that will be needed by the sales and marketing organizations are evaluated, it's important to review the methods that will be used. In Chapter 5, we described these as follows:

- Identifying, documenting, and prioritizing customer needs
- Analyzing market and competitor data
- Incorporating findings into the company's strategic plans

To identify the needs of the customer, the sales and marketing organizations must identify and closely monitor the customer base. New markets and potential markets will need to be discovered and evaluated. Then those markets must be penetrated. New generations of extremely lightweight and portable computer and telecommunications equipment are proving to be invaluable in locating and assessing these markets.

Identifying new markets and potential markets through on-line data bases will

The machines used in obtaining business.

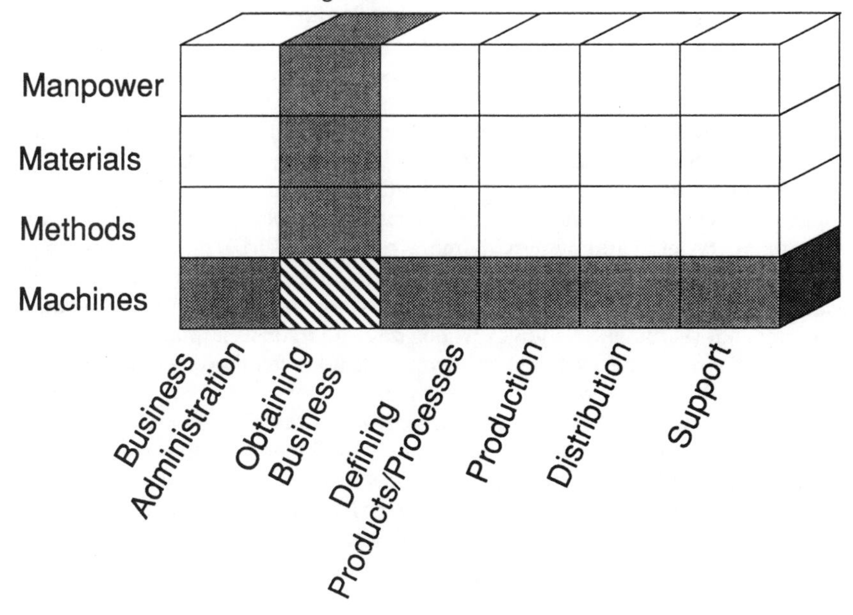

require a PC and a modem. They may be monitored in some respects through these on-line services as well. In addition, actual sales numbers, communicated from point-of-sale transaction-tracking devices, will be more closely monitored. Many requests for product specifications, features, and prices will be handled on-line in the future (a few already are, but not generally in mainstream manufacturing companies just yet). The information may be sent via E-Mail or fax. Potential markets will be uncovered by scanning sales and other indicator (leading and lagging) data, mapping trends, and watching for correlations.

Again, many of these functions can be handled via PC. Other, larger-scale activities that involve computation beyond the capacity of a PC, or cases where it is more efficient to share the raw data, can be handled by minicomputers or mainframes at the facility or at some remote marketing office, with summarized findings telecommunicated to portable equipment as they become available. These activities will require significant investments in voice communications and digital communications controllers, and in general EDI equipment.

Analyzing market and competitor data at any significant level is generally the purview of mainframe computers. The kind of number crunching required to analyze a sea of data and turn it into meaningful sales projections, market profiles, and other sound advice is substantial. As more information becomes available, the job will get bigger, requiring faster and "smarter" computing and communications gear. EDI equipment will be extremely important in this environment, with cellular telephones and fax machines becoming prolific as they get smaller, more portable, more durable, and less costly. The cataloging and management of customer requirements will become a far more important element of the marketing and sale profession, particularly in the higher-ticket end of manufactured goods. Generating and maintaining a complete and accurate profile of customer needs will become a major aspect of competitiveness, and computers offer a unique capability in this area. Laptop computers are already being outfitted to maintain product configurations. With this tool, sales professionals can configure the product (be it automobile or airplane) to the customer's specifications while sitting with the customer, never offering option combinations that are not possible, and show the customer in full color exactly what the finished product will look like. In addition, with the application of EDI technology, the order can be placed on-line, exact prices can be quoted, and delivery dates can be projected from on-hand inventory and system-held manufacturing and distribution cycle times.

As market and sales analyses encompass larger volumes of data, two things will happen: faster equipment with more memory will be needed, and expert systems will have to be developed to perform gross analytical functions. This means that a greater percentage of companies' mainframe computing resources and systems development resources will be devoted to obtaining business. Innovative ways of structuring systems architecture will be required, including mechanisms for handling the masses of data at mainframe level, passing critical details and "intelligently" summarized data to department-specific systems, and isolating appropriate data subsets with "intelligently" developed, system-generated "advice" to remote and office-located sales support devices.

Systems architecture will also have to work in reverse, allowing collection and appropriate analysis/summary of the data fed back into the system from the aforementioned and other remote (e.g., point-of-sale) devices. Those products sold to localized markets, or those where sales professionals are locally constrained, will likely foster the widespread application of metropolitan area networks (MANs), which interface with corporate wide area networks (WANs).

The Machines Used to Define Products/Processes

The design of machines utilized by mainstream manufacturers over the next decade to define their products and processes will be heavily influenced by the current movement toward concurrent engineering. The three-dimensional and solid model design capabilities that will become the norm over the next decade will require mainframe-level workstations (generally Unix operating systems with Oracle or equivalent data base management systems) and some minisupercomputers (again probably Unix with Oracle or equivalent).

As we plan our strategies for information technology in design, it will be important to keep several things in mind:

- Less than 20 percent of the workweek of engineers is spent actually doing designs.
- Activities that comprise much of the balance of engineers' time include memo writing, work scheduling, and standards identification and review.
- System-based productivity improvements in this area will most likely be achieved by combining the CAD/CAM functions of current design systems with such functions as word processing, simulation, electronic data interchange, and desktop publishing.

The machines used in defining products/processes.

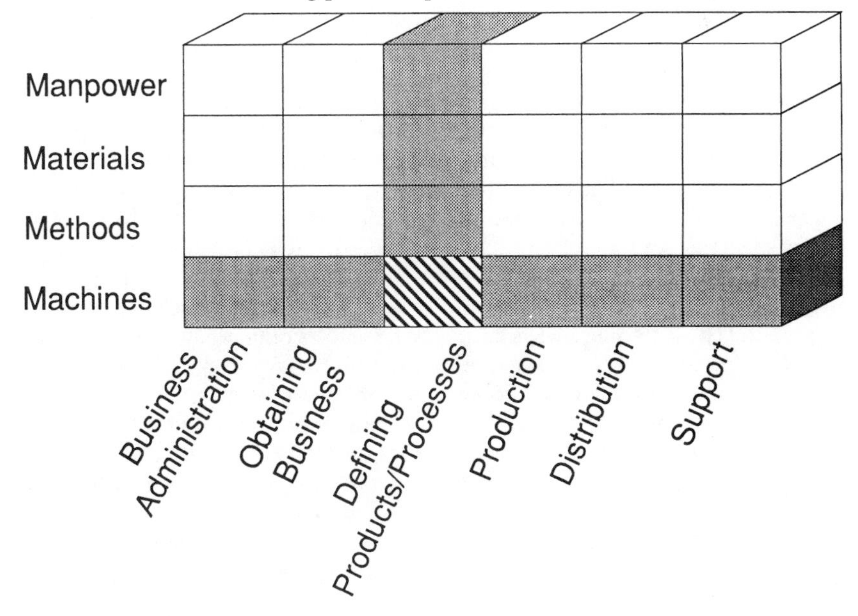

Before the end of this decade, standard features of concurrent engineering workstations will almost certainly include the transmission and publication of documents that include text, numerical data, and both 2-D and 3-D images. Animation capabilities will likely be included in many systems as well, which should prove especially useful in process simulation work.

Toward the end of the decade, as concurrent engineering begins to reach its potential in mainstream facilities, LANs will be used to provide for simultaneous review, analysis, and manipulation of product and process designs by multiple people in multiple disciplines. Generally, the design engineer's workstation will serve as the hub of this activity, with multiple (often much less powerful) terminals connected via LAN in other areas of the facility, or in entirely remote locations. These terminals will be used by manufacturing engineers to develop manufacturing process routings, by procurement people to initiate supplier information (e.g., quotes), by product and customer support professionals to perform logistics and replacement part analyses, and so on.

Coping With the Information Onslaught

Like the sales and marketing data described earlier, the information analyzed during the concurrent engineering process will grow in both volume and complexity throughout the foreseeable future. We will find ourselves maintaining data bases of process capability, materials properties, part families, and process control performance as well as other information not widely utilized previously in the design process.

All of this information must be gathered, stored, analyzed/summarized, and communicated as required during the design process. New devices will be needed to collect this information (in many cases real-time from the manufacturing floor), aggregate it, and store it for subsequent use. Existing devices will be modified to be more robust, more durable, and faster. They will become standardized in terms of output protocol and attachment/application to factory equipment. They will become much smaller and less expensive.

In the first decade of the next century, as nanotechnology begins to appear on the manufacturing horizon, extensive simulation capabilities will be required to model complex chemical compounds and evaluate their properties. Constructs must then be developed incorporating the newly designed materials into parts and developing new processes or modifying existing processes to produce the new materials and parts.

Eventually, the simulations and processes developed on the computer will be converted into manufacturing instruction sets downloadable directly to production equipment. (These developments are probably fifteen years away at the elementary manufacturing level.) When these developments appear, they will mean that we have found ways to miniaturize robotic assembly equipment to the extent that it can handle material in very small "chunks" (a few molecules) at a time. We will also have found ways to attach sophisticated (and also miniaturized) sensor and observation equipment to the robots, and ways to transmit instructions and process control feedback data between technicians' computer terminals and the miniaturized equipment.

Leading-edge instrumentation is already in limited use. Among these new pieces of equipment are the scanning tunneling microscope; double-alignment ion-scattering equipment; high-resolution electron loss spectroscopy equipment, angle-resolved photoemission equipment, which utilizes synchrotron radiation; auger spectrography tools; low-energy electron diffraction gear; electron microscopes; field ion microscopes; and atomic probes. In terms of product and process definition, this truly is, as best we can tell today, the "final frontier."

Machines Used in Production

Major changes anticipated over the next decade in manufacturing equipment are expected to support significant progress in manufacturing efficiency, applications of new materials, and shortened cycle-to-market. Consider this excerpt from *Strategic Technologies for the Army of the Twenty-First Century*, by the National Research Council (National Academy Press, 1992):

> Technological advances in material transformation processes combine new scientific understanding of the underlying transformations with automated control systems to monitor and control the process. These changes in processing technology will accelerate three trends:
>
> (1) the ability to specify the attributes of a material ("designer materials") will broaden to include the ability to design and fabricate "designer parts"; (2) the information subsystems component of larger systems will increase; and (3) the reproducibility of processes and control information will increase the ability to model variations in process variables and predict system performance.

The machines used in production.

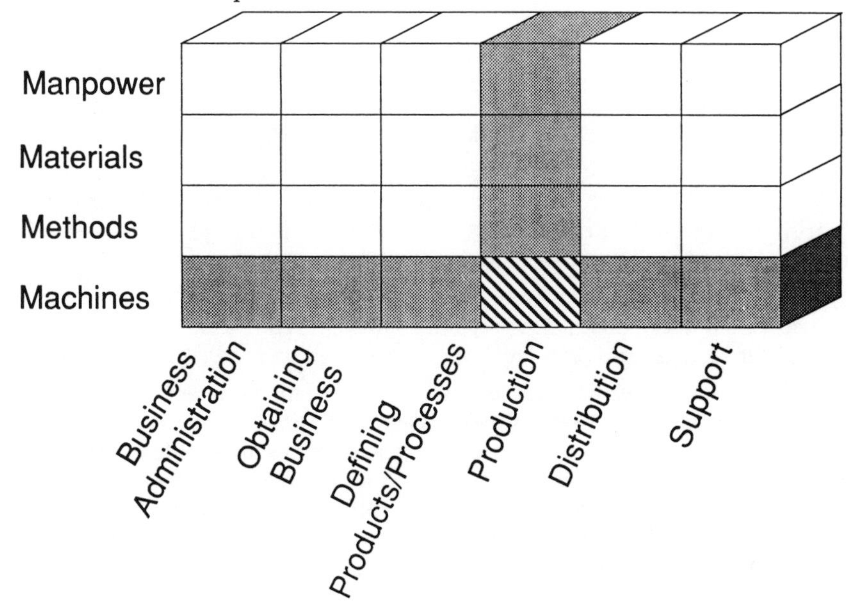

These changes will require that much more sophisticated sensors and monitors be developed, and that these improved instruments be merged with advances in the field of robotics. Intelligent control systems will also be required as the data volume and management of more precise elements (supporting microfabrication) becomes the norm. Eventually (probably in the first decade of the next century), we will move closer to microfabrication and, in the following decade, nanofabrication. The critical path technologies enabling these developments will include the sensor, the intelligent controller, and the robotics miniaturization and precision advances mentioned in the previous section.

In the nearer term, however, the majority of mainstream manufacturers will focus on flexible manufacturing systems (FMSs), the application of expert systems to individual equipment and FMS work "cells," the development of process control devices, and their application to individual pieces of manufacturing equipment. Robotics applications in both fabrication and assembly environments will continue to accelerate, with increased precision and sensitivity in end-effectors (the "hands" or "fingers" of a robot), greater resistance to hostile and widely varying work environments, and continued system cost reduction being primary areas of concentration.

Precision and control (repeatability) are among the greatest enablers of product quality. Quality will continue to be the most important distinguisher of products and services and, as discussed in Chapter 2, the key to added market share. Therefore, production equipment development resources will increasingly be focused on these areas.

Ten Trends in Production Equipment

Other, less dominant but still significant trends in the evolution of production machinery will include these ten:

1. General increases in the application of automation. Clearest examples include the addition of automatic tool-changing devices, automatic work-loading devices, operating condition sensing/reporting devices, and measurement and calibration devices to existing numerically controlled (NC) machines.

2. Increased application of nontraditional cutting and finishing process equipment to fabrication machine centers, particularly in the areas of gear cutting, shearing, and thermal cutting. The most popular near-term applications will utilize lasers and electrical discharge machining (EDM) equipment.

3. Advanced coating equipment will be developed and widely utilized to treat a variety of surfaces, most notably tooling surfaces. Treated tools will last longer, remain sharp longer, and offer improved cutting ability in newer, harder materials.

4. Fixture advances will multiply as the result of increased effort and resource expenditure over the next several years. Flexible fixtures will be developed. They will eventually be assembled, altered, and disassembled by robots in many cases to minimize time, human labor, and variation from use to use. Several approaches to fixture problems are in development, spanning the range from mechanical, fixed-position "hard fixtures" to fluid-based solutions. Clamps, vacuums, molten quick-cast, and recyclable fixtures are other approaches being studied.

5. General sensor technology development will be accelerated. Micromechanical sensors, three-dimensional vision systems, "artificial skin" heat and tactile sensors, and other special-purpose sensors are under development and will be used in much wider applications.

6. Application of "intelligence" to robotic devices will generate "smart robots" with built-in decision-making ability. These devices will eventually be able to "read" sensory input from their tactile, thermal, and visual sensors and make decisions about how and when to perform what functions on the basis of that input.

7. Application of fiber optics in production and maintenance equipment.

8. Embedding of diagnostic information and expert system decision-making capability in production and maintenance equipment.

9. Application of natural language processors to production and maintenance equipment.

10. Application of 3-D solid model displays on-line on the production floor, replacing current paper drawings, manufacturing process planning, and instructions.

The Challenges We Face

Several specific challenges to the application of new materials in production must be overcome through equipment and process development in the next decade, including the following:

- Maintaining the stability of microstructure and interfaces, particularly in the fabrication processes used in polymer-based composite materials.
- Improving control and predictability of interfacial structural and chemical change during temperature-critical and time-critical production processes, such as autoclave cure cycles.
- Improving the accuracy of process measurement and the flexibility of process control devices.
- Improving and simplifying techniques for technical cost modeling so that they consider not only the economic and performance trade-offs of alternative materials but also the respective costs associated with different processing methods and equipment. Effectively combining the findings of both analyses will be essential to making intelligent decisions about the specific equipment to be developed or purchased for future production operations.

Perhaps the most important thing to remember about design equipment over the next decade will be its role as the medium for integrating the activity of multiple disciplines within the company. It must become flexible enough and powerful enough to take in market and supplier data, incorporate the work of multiple internal organizations, and support the intellectual intercourse of product and process design integration simultaneously.

The Machines Used in Distribution

The machines utilized in distribution over the next decade can be considered in three broad categories: packaging equipment, material-handling equipment, and transportation equipment. The most fundamental and visible changes are in the first two categories.

Packaging will increasingly be tailored to individual products and product-shipping configurations. Product lines will become more diverse to handle more diversity and specificity in customer requirements. Therefore, packaging methods and equipment will have to be more flexible than ever before. Fortunately, computerization (especially computer-based design systems such as CAD, CATIA, and UNIGRAPHICS) is converging with robotics, new packaging materials, and simulation technologies to achieve that flexibility. Computer modeling is being used to analyze packaging configuration alternatives at an accelerating pace as we enter the mid-1990s. These modeling capabilities, while still embryonic, will eventually be readily available in system packages that analyze and optimize packaging methods, materials, and labor costs to identify the least-cost and highest-quality alternatives.

With distribution professionals' active participation in the concurrent engineering process becoming prevalent toward the end of the 1990s, this approach will become extremely effective. Later (in the following decade), as knowledge-based systems are employed in the packaging analysis and selection systems, they will become even more efficient than effective. Toward the end of the 1990s, the computer-based packaging specifications selected will be communicated directly from the packaging model to the machine centers that fabricate and assemble the

The machines used in distribution.

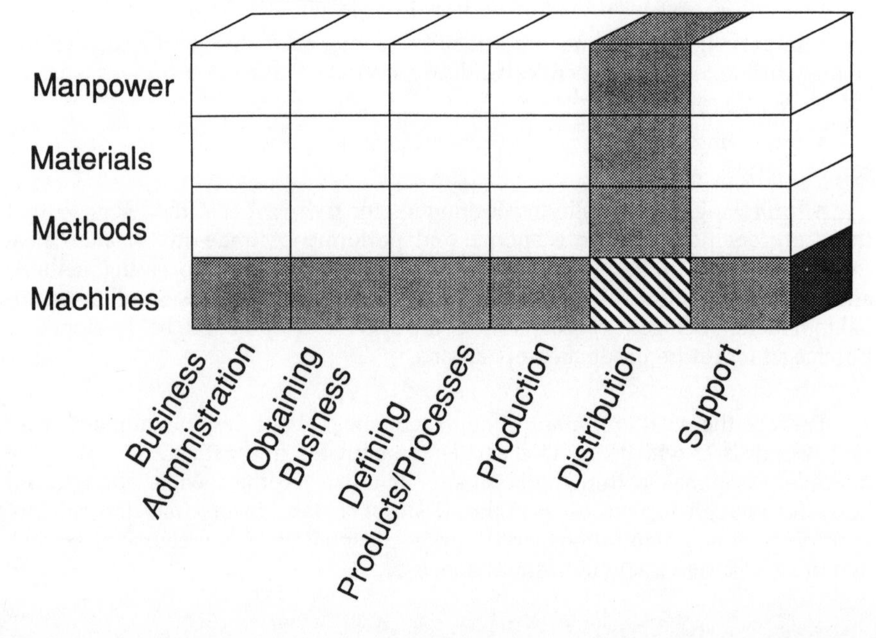

packages. This will be easiest when package fabrication facilities are in-house, but EDI will mitigate a great deal of that distinction as time goes on.

Stumbling Blocks in the Shipping Room

With the application of robotics, both packaging fabrication and assembly and material packaging activities will be performed by machine. Some of this occurs today, but there are serious limitations caused by the struggle to standardize package sizes to most efficiently utilize warehousing, racking, conveyor, handling, and transportation equipment.

Material-handling machinery developments may be traced to the convergence of several technologies, including robotics, flexible manufacturing systems, automatic test equipment, sensor technology, and automation control systems. However, until concurrent engineering took hold, few manufacturers considered packaging cost and effectiveness in the design of their products and processes. Almost none considered the unique requirements of automation applications in this process.

Many tried to automate (spending a lot of money for very little efficiency improvement) without thinking adequately through the ROI aspects of their projects. They became enamored with the automation. In so doing, management was disappointed and the prospects for future automation were seriously impaired. Through the 1990s, most of the growth in automated material-handling system automation is expected to be in small- and medium-scale automation systems, with most companies developing and implementing their automation one piece at a time. Experts in this area expect applications in general automated material-handling systems to double by the end of the 1990s.

Letting "Travelpilot" Do the Navigating

Transportation equipment will continue to be affected by the development and application of computers as well, but the changes will be a bit more subtle. One example is a vehicle-based navigation system called "Travelpilot" by Etak. Travelpilot utilizes German computer hardware and CD-ROM maps to continuously monitor and display a vehicle's position to the driver as it travels. Maps are currently available covering about half of the United States, and more cities will be added. The vehicle is depicted as an arrow superimposed on a scrolling map. Destinations may be selected and marked, and arrows will then direct the driver toward the specified destination. As of the time of this writing, the Travelpilot costs about $2,500, with maps other than the one (presumably for your area) originally purchased costing $150 each.

Like all other computer-based technology, the price of such systems is likely to come down even as product sophistication increases.

Machines Used in Product and Customer Support

In Chapter 7 the machines currently used in customer and product support were broken down into four broad categories: communication equipment, customer- and product-tracking equipment, training equipment, and product service equipment.

The machines used in customer and product support.

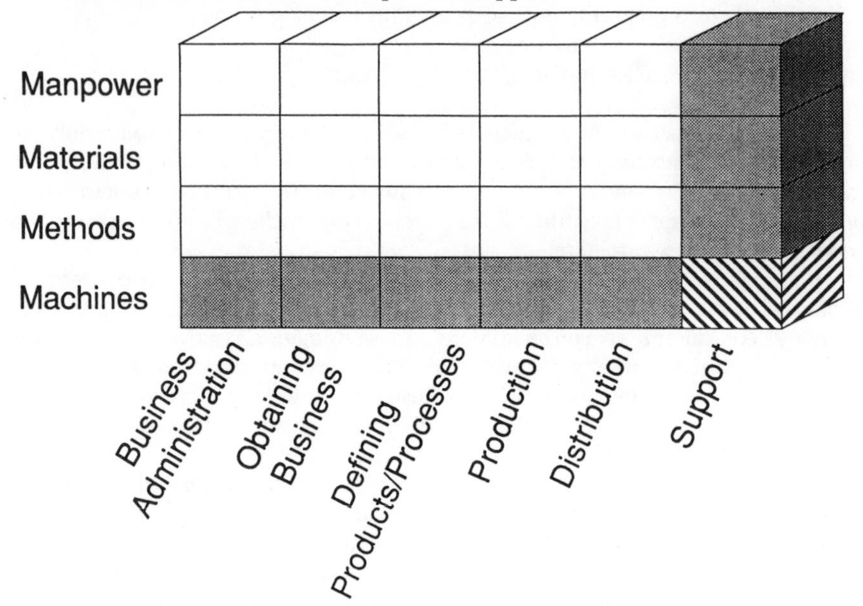

Breakthroughs in Communications Equipment

Communications equipment includes telecommunications gear such as telephones and modems as well as equipment used to communicate in print, such as service manuals and technical publications. In this area, the merging of communications media will be a prominent feature affecting equipment design and deployment. More sophisticated equipment will appear that combines visual, audio, graphic, and textual information for owner/operator manuals, service information, and warranty statements. The information will be carried on CD-ROM or similar media.

Anyone who has ever attempted to assemble a bicycle or other consumer product marked "assembly required" will recognize this fact: In most cases, we don't look at the instructions until we've failed to get it right on our own; in many cases, we only look at the "exploded view" drawings and captions, and only when all else fails do we actually read the full-blown assembly instructions.

A picture truly is worth a thousand words. Even *more* powerful than a picture is a videotape of someone actually performing the assembly, and *still more* powerful is the combination of video and sound with instructions given verbally as they are demonstrated visually. By the end of the 1990s, with the prices of VCR tapes and newer technology such as CDs driven to very low levels, it seems likely that this communication medium will replace the printed instructions usually provided today.

In terms of actual service equipment, the types of communication equipment provided are likely to build on these technologies and produce some form of interactive video. Interactive video in this context will be made possible through the convergence of these new audio and video technologies with expert systems, natural language processors, and other artificial intelligence techniques. These interactive videos will provide the capabilities of an expert system with the user-friendliness of an answering machine or television. When the service repair operations prove too complex for the buyer, the fiber optic cable will come into play. It may well utilize devices such as the new "picturephone" (if it catches on this time), allowing factory technicians to help users directly in their own homes without leaving the factory.

Prior to the emergence of these interactive videos, diagnostic intelligence in the form of product-embedded computer chips is likely to be an increasingly common device. This approach will reduce actual service time and reduce the number of returns to service centers for maintenance.

Today, for example, malfunctions in some equipment can be corrected by telephoning a factory-based service center that can hook directly into the malfunctioning equipment by telephone. Over the phone, technicians can diagnose the problem and correct it. This phenomenon will become more common as the level of "intelligence" within products continues to increase.

What Our Products Can Tell Us

In terms of product- and customer-tracking equipment, the increasing level of embedded intelligence will also prove extremely useful. Simple devices are generally built into the least-expensive equipment, with the converse being true as well. One very simple example of product-tracking equipment is the theft-deterrent device often attached to clothing and audiotapes. This device alerts the retailer when merchandise is being removed from the store before the device is deactivated. At the other end of the spectrum is the monitoring equipment built into space shuttles, which provides "real time" feedback on hundreds of variables to earthbound scientists while recording many more for later study upon the shuttle's return.

Increasingly, these devices will be used not so much to *determine* the product's performance characteristics as to *verify* what was anticipated to happen during product performance simulations. Determining product performance characteristics will still be a vital function, but simulations will become more important as customer and product support professionals become more active participants in concurrent engineering activity.

Therefore, an important design characteristic of these devices will be the ability to effectively communicate, aggregate, and utilize the information by incorporating it into new and revised product designs. How the devices and the equipment used to collect data from the devices interface will be critical, and will likely be the subject of serious R&D expenditures in this field, beginning around the year 2000.

When Virtual Reality Becomes a Reality

It is important to reiterate here that eventually the technologies packaged together under the term *virtual reality* will have a substantial impact on the machines used in product and customer support. The equipment and software encompassed in these systems will enable service technicians to perform virtually any repair with very little real "training." John Walker, in *Through the Looking Glass* (Addison-Wesley, 1990), states:

> Users struggling to comprehend three dimensional designs from multiple views, shaded pictures, or animation will have no difficulty comprehending or hesitation to adopt a technology that lets them pick up a part and rotate it to understand its shape, fly through a complex design like Superman, or form parts by using tools and see the results immediately.

Current equipment used to support this technology includes devices like "gaze-trackers," projectors, force-feedback apparatus, sensing gloves, goggles, and telerobotic controllers costing millions of dollars. However, as these costs come down (and they already are), the devices will be widely deployed. We are likely to see effective use of this technology sometime during the first decade of the next century.

Chapter 15

Using the Model to Identify Your "Critical Path" and Manage Change in the Twenty-First Century

In the book *Running American Business* by Robert Boyden Lamb (New York: Basic Books, 1987), the author states, "Trying to anticipate . . . a new, broad-based technology is like trying to capture lightning in a bottle." Nonetheless, this kind of projection is precisely what we *must* learn to do in order to survive and succeed in the years ahead. The information and the model offered in this book provide a useful tool for exactly that purpose.

The model introduced in Chapter 1 and developed in Chapter 8 can now be employed to help the reader anticipate technological change. Putting the information and trends reviewed in previous chapters into the model and adding the unique elements and environmental factors from the reader's specific industry, we can gain insights about how that industry will evolve and what specific strategies can be deployed to best meet the challenges ahead. What specific steps should be taken in what order can be regarded as the reader's "critical path."

If, while deploying our resources on jobs along other paths, we fail to constantly apply resources to the critical path operations, the entire product cycle will be extended. It will take longer overall to accomplish our objectives, and we will waste time and money. In the meantime, our competitor may be quicker to market and put us out of business.

Consider the job of navigating a manufacturing company through the sea of changes that will surround and permeate us over the next two decades. Only by analyzing the business in these terms can we hope to manage our resources effectively enough to survive and compete. Only by identifying and applying innovation to the "critical path" of our own development can we win.

To understand this concept, consider an actual manufacturing critical path: The critical path in an assembly operation is the longest string of jobs that must be performed to complete the assembly. For example, in Exhibit 15-1, the critical path must be identified from all the possible paths, as follows:

A. #3 to #6 to #9, which totals 8 hours
B. #52 to #6 to #9, which totals 5 hours
C. #12 to #31 to #9, which totals 13 hours
D. #62 to #9, which totals 7 hours

Exhibit 15-1. Identifying the critical path.

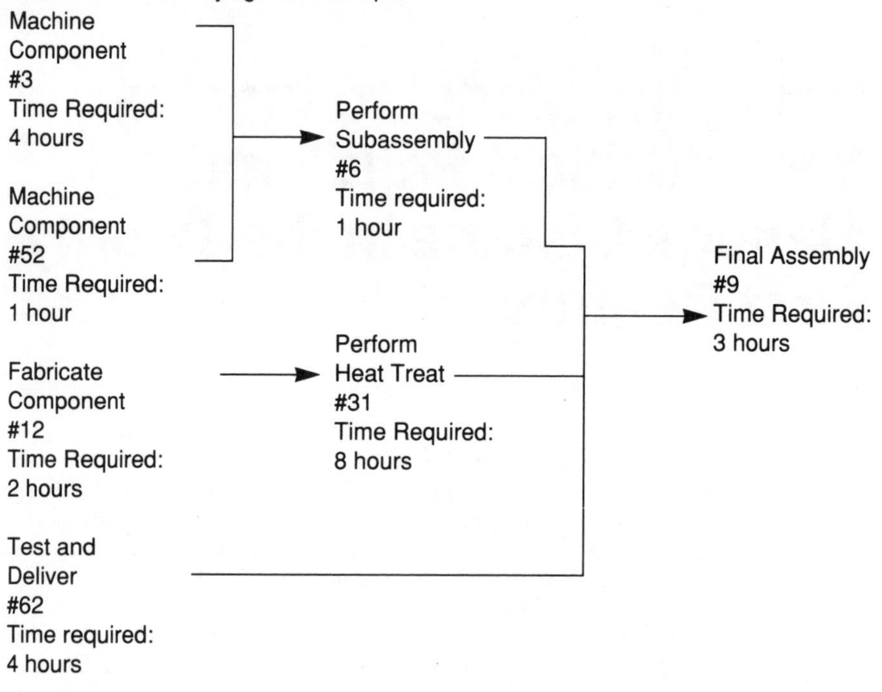

In this very simplified example, path C is the critical path, because it is the path requiring the longest time to achieve the completion and delivery of the assembly. Constantly manning the critical path will ensure that the product is produced in the least possible time.

How the Model Can Help Us Find the Critical Path

As manufacturing leaders and stakeholders, our goal is to reach the "future state" we envision for our manufacturing company. There are hundreds of paths that must be explored to take us there, including anticipated developments in every cell of our model, every area of our business. The model we have been referring to through the course of this book offers not only ways to identify potential developments in each of these areas but also a structure for identifying how these developments should be ordered. It can help us to determine what ought to be done first, second, and so on, and what the critical path is, so that we can keep it constantly at the forefront of our attention.

The most revealing insights from this model can be gained by careful analysis of all model elements and how they may converge. These are the most useful exercises because they provide glimpses of what the future can be, given not only objective evidence but the imagination and insight of the modeler as well. These insights are remarkable springboards for new product development, new service offerings, and the development of new business processes. Each picture will be unique because each business is unique, and the individual talents and knowledge of different people will be applied during the modeling process.

However, there are some principles governing the dynamic interactions that occur over time that are consistent from company to company. When these principles are understood, they offer important indications about where companies could deploy their R&D resources most effectively. We will briefly consider the most basic principles here.

Basic Principles of Model Analysis

First of all, as just mentioned, each company is different. Most have organizational structures and reporting relationships that are hierarchical in nature, and most contain the primary business processes reflected in our model. However, all companies have idiosyncrasies in terms of their processes and organizational relationships.

Therefore, in order to apply these principles, the model needs to be altered to reflect the unique configuration and relationships of specific companies. For these reasons, the principles discussed here can only be regarded as the beginning point for this type of analysis in your particular company. The results of your predictions will be only as good as your model. With these caveats, then, we will review the principles and elements of interaction analysis and prediction.

Consider the model as a cube-shaped, three-dimensional universe where the elements in each subcube (or cell) have a chemical "valence." That valence causes the elements to respond differently to each of the other elements, and to different combinations of elements. Like chemicals, the "ionic charges" and other properties of these elements will vary in strength. They may (or may not) also lose some of their properties (or experience "fading") over time. For examples of how the elements interact, refer to the discussion in Chapter 8 about convergence, which mentions the effects of combining elements like laser welding with other elements such as fiber optics.

When a new element appears (often as a result of the convergence of other previously existing elements), it is important to note its initial point of incursion to our model universe. For example, in the case of artificial intelligence (AI), this element first appeared significantly in the "Defining Products/Processes—Materials" cell of our model. That cell, then, is the "point of incursion" of AI.

When we have identified the point of incursion, it would help us to know where the next logical application of this new element is, so we could "go with the flow," so to speak, minimizing resistance and capitalizing on the natural evolution of the new element throughout the organization. The question that requires an answer is, Where is the most receptive position for continued application of this new element or technology? In other words, Where am I likely to achieve the least-cost growth of this technology in my company? or, Where am I most likely to see this spring up next? The answer to this question lies to a large extent in the relationships and organizational flows of the company.

For the most part, because of the internal "supplier/customer" relationships within the company, new elements (especially new technology applications) tend to travel downstream from supplier to customer. *Downstream* in our model universe generally flows from left to right. So the first principle is that, in general, the path of least resistance will be reflected in our model by movement from the initial point of incursion toward the downstream (further right) cells of the model.

If we use our AI example, this means that an extremely high probability exists that, within the decade, an application of AI will evolve in the cell known as "Production—Materials."

Equally likely is the application of the new element or technology in the same cell of our model in the following decade. In other words, there is also an extremely high probability that other applications of the new element will be found in the same cell over the next decade. In our example, this means that more than a decade later, new applications for AI will still be evolving in the "Defining Products/Processes—Materials" cell.

The next most probable cells to experience an AI application are the balance of the cells of the same business process in the same decade (the vertical slice of our model). So, in our example, the next most probable areas of AI applications are as follows:

- "Defining Products/Processes—Manpower"
- "Defining Products/Processes—Methods"
- "Defining Products/Processes—Machines"

This implies that a manager who experienced successful application of AI "material" (software) may find ways to apply it to handling manpower challenges (e.g., selecting appropriate design engineers to lead a new design project), methods challenges (e.g., knowledge-based systems for decision support of stress engineering activity), or machine challenges (e.g., using an expert system to determine optimum areas to be depicted in "interface control drawings").

Other cells of the model that are also very probable locations of comparatively near-term application given the point of initial incursion are the cell immediately upstream (left) of the point-of-incursion cell in the same decade, and the same-business-process cells (the vertical slice) in the following decade. Again returning to our AI example, the upstream cell would be the "Obtaining Business—Materials" cell.

The same-business-process cells in the next decade are the following vertical slice cells:

- Defining Products/Processes—Manpower (following decade)
- Defining Products/Processes—Methods (following decade)
- Defining Products/Processes—Machines (following decade)

The term *decade* as used here is a guideline, not a hard-and-fast rule of ten years. More aggressive change implementers will experience shorter intervals, and those in more recalcitrant environments will take longer to effect change. In addition, within the same company there will be higher and lower levels of "receptivity" from cell to cell depending on the people and processes involved. These overall receptivity changes determine to a large extent the "valence" of the elements in that cell, and consequently the ability of the new element to find a home there.

For example, smaller companies tend to be more closely knit and therefore quicker in moving new elements between cells, in developing new applications. They have a higher propensity for innovation and so are able to identify and exploit new applications over a broader spectrum of the organization in compressed time frames. In these companies, the term *decade* could perhaps better be replaced by another less explicit term, such as *generation*.

Working from the foregoing principles and similar rules, I developed a table of general incursion receptivity that may be used to describe the receptivity values of other elements according to spatial relationships between the cells those elements occupy and the initial point of incursion (see Exhibit 15-2).

When depicted graphically, the concept becomes somewhat clearer. The values in the cells represent the general receptivity ratings of the elements in those cells to new applications and resulting "chemical" change from the new element, given its point of initial incursion. If we look at our example of the incursion of artificial intelligence into the manufacturing company we see a picture that looks like that shown in Exhibits 15-3 and 15-4.

Now the tricky part. There are a few formulas that may be used to analyze the likelihood of new element evolution into cells and "chemical" reaction between the newly introduced element with the elements of the other cells. They include the incursion probability calculation shown below:

$$IP = (IC \times RV) \times 10$$

where

IP = Incursion Probability
IC = Ionic Charge
RV = Receptivity Value

In support of the incursion probability calculation, there is a series of subcalculations required:

$$IC = PW \times PT \times PR$$

where

PW = Power (growth rate)
PT = Potential (potential migration estimate)
PR = Pervasiveness (migration rate analysis)

and

$$RV = (IO/IE) \times PF$$

where

IO = Intracell Occupancy (the number of elements within a cell that have a perceivable application)
IE = Intracell Elements (total number of elements within the cell)
PF = Proximity Factor (factor determined by proximity of target cell to the cell that was the initial point of incursion)

Exhibit 15-2. Proximity/receptivity factor table.

Receptivity Value	*Relative Cell Location*
10	Direct interface cell downstream in same decade
10	All elements within the same cell
10	Same cell in following decade
9	All vertical cells (same-business-process cells) in same decade
8	All vertical cells (same-business-process cells) in the following decade
8	Direct interface cell upstream in the same decade
8	Direct interface cell downstream in the following decade
6	Direct interface cell upstream in the following decade
5	All noninterface downstream cells in the following decade
3	All noninterface downstream cells in the same decade
3	All noninterface upstream cells in the following decade

Exhibit 15-3. Current decade.

	Business Administration	Obtaining Business	Defining Products/Processes	Production	Distribution	Support
Manpower	1	1	9	3	3	3
Materials	1	8	10	10	3	3
Methods	1	1	9	3	3	3
Machines	1	1	9	3	3	3

= Point of Initial Incursion

Exhibit 15-4. Following decade.

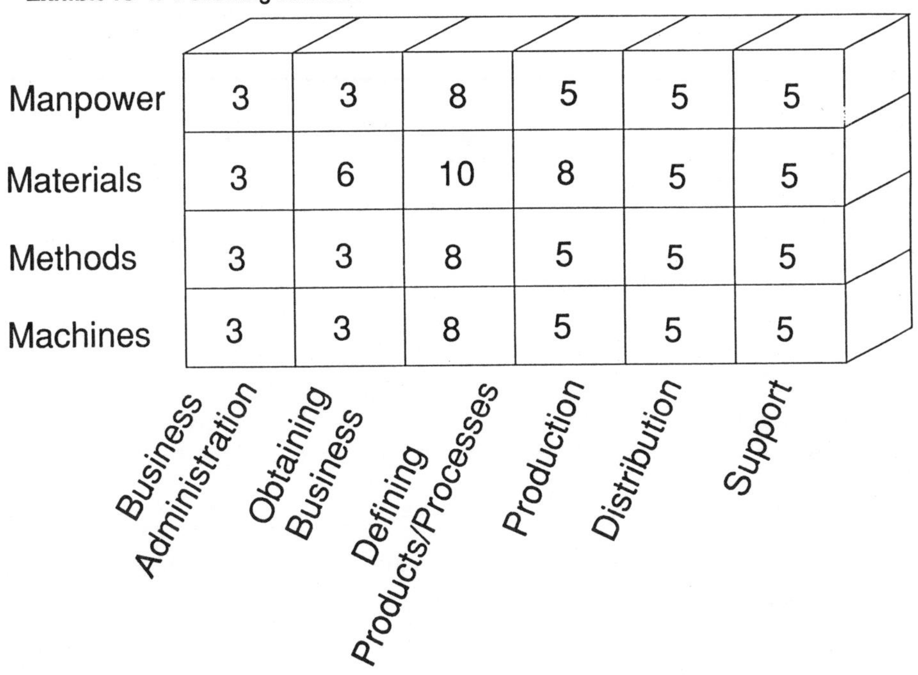

	Business Administration	Obtaining Business	Defining Products/Processes	Production	Distribution	Support
Manpower	3	3	8	5	5	5
Materials	3	6	10	8	5	5
Methods	3	3	8	5	5	5
Machines	3	3	8	5	5	5

To perform the growth rate analysis, the growth in sales of the elemental technology should be used when that is available, normalizing the growth over the period of months between three distinct measurement points in time. For example, the growth rate analysis for AI in "Defining Products/Processes—Materials" might be estimated by identifying the published market size for this software (measured at $35 million, $130 million, and $900 million over a span of 60 months).

Determining the cell migration rate (PR) is a matter of identifying how many cells this element occupied over this same 60-month period, measured at the same intervals. In our AI example, this might have been one cell in period 1, four cells in period 2, and nine cells in period 3.

Arriving at the potential migration estimate (PT) involves identifying the number of cells with primary application potential (somewhat subjective, I grant you) and the number of cells with secondary application potential. In our AI example, I estimated these values to be 2 and 6, respectively.

These analyses would produce a "power" (PW) rating of .72, a "potential" (PT) rating of .12, and a "pervasiveness" (PR) rating of .35 over a ten-year span. Plugging these numbers into our formulas, then, might yield the following analysis of the probability of AI applications in "Distribution—Methods" in the following decade:

$$RV = (IO/IE) \times PF$$
or
$$RV = (5/12) \times 5$$
or
$$RV = 2.08$$

$$IC = PW \times PT \times PR$$
or
$$IC = .72 \times .12 \times .35$$
or
$$IC = .03$$

Therefore, the incursion probability, IP, can be determined thus:

$$IP = (IC \times RV) \times 10$$
or
$$IP = (.03 \times 2.08) \times 10$$
or
$$IP = .62$$

In other words, there is a 62 percent probability that an application of AI will develop in that cell within the following decade. (I suspect that the scientific community would be embarrassed at the liberties I have taken in this approach, but it works for me.)

Using this approach, strategic planners can generate powerful visions of the future state of their operations and better understand the temporal relationships involved in realizing those visions. When the fabric of these relationships has been exposed, the planner can more readily identify the sequence of technological and physical developments that must occur to move from the current state of operations to the future state, and the likelihood of each major development over time. It will become much clearer to planners where development resources may be most effectively applied in order to constantly work the "critical path," which is what companies must do to reach their strategic goals and most effectively utilize their resources. It is what we must do as American manufacturers to win in global competition.

What to Expect Along the Critical Path

Because every industry (and every company within each industry) is different, each company's strategic vision will be different. As a result, the critical path for implementing and managing change will be unique in each company. Some jobs are likely to fall on the critical paths of most manufacturing companies, however, because of their importance in terms of the way manufacturing as a business will change over the next twenty years. Among them are the following:

- *Design systems,* which will expand in scope to cover strategic planning elements and tactical development initiatives of most company disciplines, under the auspices of concurrent engineering. In this area, knowledge-based systems will serve to provide tremendous leverage of vital human "intelligence" in simulated form through expert system applications. This development will be foundational in achieving meaningful gains in cycle-to-market—one of the most critical differentiators of successful companies over the next decade. The critical path in this area will likely include development of individual expert system modules, improved module interfaces, improvements in the flexibility, interfaces, and responsiveness of artificial intelligence systems, and advances in computer simulation (both hardware and software).

- *Fabrication systems,* which will evolve from brute-force bending, shaping, and cutting of metals in many cases toward formulating, mixing, curing, and joining of polymers, ceramics, and glass. These kinds of changes will require different equipment, different training, and different raw materials from those we have traditionally used in American manufacturing (with the notable exception of the chemical industry, which is no small clue about who may be in the strongest position to profit in these areas).
- *Assembly systems,* where immediate gains will most likely be focused on reducing the need for mechanical fastening. Longer-term critical path elements will include strategic changes in the location of final assembly operations and the development of advanced, multimedia training delivery systems.
- *Infrastructure management systems,* which will become increasingly critical in the regulation-intense and data-rich environment of the 1990s and beyond. This will involve dramatic systems development and integration efforts that integrate internal information management with public domain systems and information flow. These changes will be critical in the most heavily regulated industries, but will be important to everyone.
- *Standards development,* especially as the standards pertain to electronic data interchange (EDI). Sound strategic decisions pertaining to which standards companies adopt will make great differences in terms of the companies' ability to market and produce internationally, and even effectively communicate with their own supplier bases.

Most of the manufacturers that do a credible job of identifying the critical path to a solid, visionary future will likely find these elements somewhere along that path.

By evaluating the propensity for change in each area (using the proximity/receptivity analysis tools described), then observing the potential outcomes of various convergences among developing elements, and finally identifying the critical path of developments that must occur to achieve the future envisioned, we can properly deploy our resources.

We can attain the future we envision. It is a matter of identifying and effectively working the critical path.

Chapter 16

Conclusions and Recommendations

Opinions, it is said, are like noses—everybody has one. Recommendations are at least as prolific. For that reason, I will not spend a lot of time on them here. However, during the presentation of this model for forecasting change and innovation in manufacturing industries, some needs have certainly been presented that bear summarizing. Here they are discussed in two broad categories: industry and government.

Industry

Manufacturers in the United States, in order to become more competitive and eventually win this most important game of our existence, must recognize and admit our problems, commit to action to resolve them, and move aggressively toward achieving the goals necessary to remedy these problems.

Awareness

In order to get better, American manufacturers must directly confront and embrace the fact that there are better ways to do things. We have to stop whining, blaming our problems on cheap foreign labor, unfair trade practices, and big government. We can't fix our problems until we recognize and admit to ourselves that they exist, they weren't all caused by somebody else, and we are the only ones who can fix them.

Action

To resolve these problems, we must commit more than lip service to them. It will cost resources in time, money, and talent. There will be a cost, but it is a cost we must pay to survive.

Orientation

We must achieve several goals to resolve our problems, including the following:

• Develop powerful visions of our companies as being the world's most successful manufacturers and enjoying a mode of growth. The visions must be detailed enough to be clear, compelling enough to motivate and inspire, and aggressive enough to ensure that the results are world-class. They must reach out beyond normal planning horizons, spanning at least twenty years. They must be customer- and future-oriented, designed to meet the future needs of the customer far in advance of the customers' own recognition of those needs.

• Ensure that the vision is common throughout the company. Every employee needs to share the vision, so that everyone is working toward common goals. All too often, regarding knowledge as power, executives hold this information at upper levels of the organization and find themselves "leading" from the rear of the "column."

• Develop complete, future-oriented strategies to achieve the vision. Whether the model discussed in this book or some other one is used to develop those strategies, the strategies must be integrated, detailed, and realistic. They must be "bought into" at every level of the organization, and funded to ensure success.

• Run the business with a bias toward growth rather than contraction. When possible, make your own components rather than buy them, particularly when the sourcing decisions involve foreign suppliers (see earlier graphics entitled "How To Go Out of Business" and "How To Grow a Business" in Chapter 10).

• Develop tight integration, among internal systems and processes first, then between your systems and those of customers and suppliers. As we have seen, opportunities to develop systemic integration across all of these fronts is unprecedented and will continue to increase over the next two decades.

How Can We Regain Our Place?

As American manufacturers, we need to recognize ourselves for what we are—the people who built the greatest economic engine that has ever existed. At the conclusion of World War II, America was absolutely incomparable. Our market was nine times larger than any other, our technology was far superior to all others, our workforce was the most highly skilled in the world, and we had the most talented and professional managers in the world.

Certainly, with the opening of the European Economic Community, our former market advantage has been erased. We are, for the first time in more than one hundred years, the world's second-largest market. Actions currently being taken to counter this trend, such as the North American Free Trade Agreement (NAFTA), will not be enough to regain it. However, as noted earlier, the size of a country's internal market is of diminishing importance. Fortunately, advances in transportation and communication technologies will continue to erode these advantages. Therefore, we may even out on this score over the next several years.

Technological superiority is an area upon which we must focus considerable attention. We are still about average in terms of our technologies with the other advanced industrialized nations of the world. We do very well in computer-based technologies and biotechnology. We do well in design and engineering tools, telecommunications equipment, automotive power trains, electronic controls, and microelectronics. We are average in the areas of advanced materials, manufacturing processes, precision machining, printing/copying equipment, and optoelectronic components. We do poorly in chip-making equipment, robotics, electronic materials (particularly ceramics), and optical storage.

We will have to invest more in our technologies, and learn to be as patient as our foreign competitors in terms of ROI. We can't afford to continue losing market share (or in some cases entire markets) to our own greed.

A second area, worker education and skill base, will be another extremely important target for us over the next two decades. We have too few engineers and scientists, too few information technology professionals, and generally too few people who specialize and excel in the fields of physics and general mathematics. This may be the most serious long-term problem facing American industry.

Our management in American manufacturing companies is no longer widely regarded to be the vanguard of the world in any area. According to one recent World Competitiveness Report mentioned by Lester Thurow in his book *Head to Head* (New York: Morrow, 1992), American managers were ranked twelfth in the world in terms of product quality, tenth in terms of on-time delivery, and tenth in after-sales service. In terms of "future-orientedness," American managers ranked twenty-second out of twenty-three countries evaluated (Hungary was the worst). Thurow also points out that, to make matters worse, Japanese managers were rated better than American managers by 48 percent of the American population.

The areas we must improve on are (fortunately for us) those areas where we can effect changes through our own efforts. Technology, education and training, and management of our own operations are all things within our control. Constructing a clear and compelling vision, developing appropriate strategic initiatives culminating in that vision, identifying the "critical path," and committing the resources required to implement the appropriate strategies can lead us back to preeminence.

Government

There can be no doubt that America needs a strong, pervasive national industrial strategy to compete successfully in the new global economy. It will require greater vision and stronger leadership than we have recently experienced. Furthermore, such a strategy is imperative.

Beyond this national industrial strategy, some individual strategies are required in several other areas as well.

Research and Development

Some time ago, a panel known as the National Critical Technology (NCT) Panel was commissioned to identify those technologies most critical to the United States and report its findings to the President and Congress. The definition of "critical" in this context was "those areas of technological development which are essential for the long-term national security and economic prosperity of the United States." Its findings were reported in 1991. Among the technologies cited were the following:

- Materials synthesis and processing
- Electronic and photonic materials

- Ceramics
- Composites
- High-performance metals and alloys
- Flexible computer-integrated manufacturing
- Intelligent processing equipment
- Applied molecular biology
- Medical technology
- Aeronautics
- Surface transportation technologies
- Energy technologies
- Pollution minimization and remediation and waste management
- Microfabrication and nanofabrication
- Systems management technologies
- Microelectronics and optoelectronics
- High-definition imaging and displays
- Sensors and signal processing
- Software
- High-performance computing and networking
- Computer simulation and modeling
- Data storage and peripherals

Almost all of these technologies are covered in discussions within this book pertaining to the future of American manufacturing. Several prestigious panels, commissions, and other organizations have recommended continued or increased federal support for basic and applied research in these areas. In fact, one such commission, The President's Commission on Industrial Competitiveness, recommended that a federal Department of Science and Technology be established to oversee and coordinate research efforts in these and related areas. It seems clear that a single, integrated directorate similar in some respects to the ones utilized by our competitors (e.g., Japan's Ministry of International Trade and Industry—MITI) will be required for the United States to continue to compete successfully for world market share.

Education and Training

Business leaders and educators have called for wholesale education reform in support of our future economic security. Individual strategies run the gamut from vouchers to magnet schools. Whatever the specific strategies, their goal must be focused on developing useful technical skills and broad, flexible reasoning capabilities that are objectively monitored and continuously improved. We must restore prestige and dignity to the office of "teacher," and fairly compensate successful teachers on a par with successful businesspeople. Finally, we must forge permanent links between the workplace and local education systems to ensure that the needs of the workplace "customer" are constantly recognized and built into the educational "products" provided by the "suppliers" of manufacturing's human resources—our schools.

Data Exchange Standards

Government needs to assume a stronger role in coordinating international data exchange standards and in coordinating domestic research efforts to avoid duplication and waste in this area.

Other Recommendations

There are myriad other recommendations that could be, and indeed have been, made pertaining to government's role in support of American manufacturing. They include policy and legislative changes in the areas of economic and trade policies, environmental and safety regulations, regulation of industry structure and competition, patent regulation, information policy, R&D support policy, and investment policy. I am neither qualified nor inclined to present my own agenda in those areas here. A good, brief overview of existing recommendations in these areas can be found on pages 131 through 156 of the book *Toward a New Era in U.S. Manufacturing* by the National Research Council (National Academy Press, 1986).

Epilogue: A Personal Note From the Author

I have been involved in manufacturing in the United States all of my professional life. I've participated as a laborer pouring molten iron, an assembler, a production scheduler in a sheet metal shop, an expediter in final assembly operations, a director of manufacturing engineers, quality assurance people, direct labor people, and materials management people. I have participated as a management consultant in the manufacturing operations of some of the world's smallest and largest companies spanning all kinds of products from candy to computers. I have also visited manufacturing operations in Japan, Europe, the Middle East, Canada, and Australia. Given this, I hope you will regard this next statement as having some credibility.

> Maintaining a strong domestic manufacturing base, especially in the new "global" economy, is vital to our national security and to our economic future. Personal experience has shown me that when it comes to manufacturing, *no one* can take anything away from America that we do not give up ourselves. God willing, we *can* recover our preeminent position as the world's dominant manufacturers, and continue to lead the world toward a more prosperous future.

Bibliography

Barker, Joel Arthur. *Discovering the Future*. Lake Elmo, Minn.: ILI Press, 1988.

Bhote, Keki R. "World Class Quality." New York: AMA Management Briefing, 1988.

Boyett, Joseph H., and Henry P. Conn. *Workplace 2000*. New York: Dutton, 1991.

Brennan, Richard P. *Levitating Trains and Kamikaze Genes*. New York: Wiley, 1990.

Burrill, G. Steven, et al. *Biotech 90: Into the Next Decade*. New York: Mary Ann Liebert, Inc. 1989.

Buzzell, Robert D., and Bradley T. Gale. *The PIMS Principles*. New York: Free Press, 1987.

Bylinsky, Gene. "U.S. Comeback in Electronics," *Fortune*, April 20, 1992.

Cappo, Joe. *Future Scope*. White Plains, N.Y.: Longman, 1990.

Cetron, Marvin, and Owen Davies. *American Renaissance*. New York: St. Martin's, 1989.

Cetron, Marvin, et al. *The Future of American Business*. New York: McGraw-Hill, 1985.

Clancy, Kevin J., and Robert S. Shulman. *The Marketing Revolution*. New York: Harper Business, 1991.

Crispell, Diane. "Workers in 2000," *American Demographics*, March 1990.

Dertouzos, Michael L., et al. *Made in America*. Cambridge, MA: MIT Press, 1989.

Drexler, K. Eric. *Engines of Creation*. New York: Doubleday, 1986.

Drexler, K. Eric, et al. *Unbounding the Future*. New York: Morrow, 1991.

Fisher, Anne. "Is Long Range Planning Worth It?" *Fortune*, April 23, 1990.

Forester, Tom. *The Materials Revolution*. Cambridge, MA: MIT Press, 1988.

Gilbreath, Robert D. *Forward Thinking*. New York: McGraw-Hill, 1987.

Gleick, James. *Chaos*. New York: Viking, 1987.

Golden, William T. *Science and Technology Advice to the President*. Elmsford, NY: Pergamon Press, 1988.

Gordon, Gil E., and Marcia M. Kelly. *Telecommuting*. Englewood Cliffs, NJ: Prentice-Hall, 1986.

Hamrin, Robert. *America's New Economy*. New York: Franklin Watts, 1988.

Hunt, V. Daniel. *Mechatronics: Japan's Newest Threat*. New York: Chapman and Hall, 1988.

Johnston, William B., and Arnold H. Packer. *Workforce 2000*. Indianapolis, Ind.: Hudson Institute, 1987.

Kaminuma, Tsuguchika, et al. *Biocomputers: The Next Generation From Japan*. New York: Chapman and Hall, 1991.

Kearns, David T., and Denis P. Doyle. *Winning the Brain Race*. San Francisco: ICS Press, 1988.

Lamb, Robert Boyden. *Running American Business*. New York: Basic Books, 1987.

Makridakis, Spyros G. *Forecasting, Planning, and Strategy*. New York: Free Press, 1990.

Maus, Rex, and Randall Allsup. *Robotics*. New York: Wiley, 1986.

Morita, Akio. "Trade Restrictions," *Fortune*, September 25, 1989.

National Research Council. *Materials Science and Engineering for the 1990s*. Washington, D.C.: National Academy Press, 1989.

National Research Council. *Star 21: Strategic Technologies for the Army*. Washington, D.C.: National Academy Press, 1992.

National Research Council. *The Competitive Edge*. Washington, D.C.: National Academy Press, 1991.

National Research Council. *Toward a New Era in U.S. Manufacturing*. Washington, D.C.: National Academy Press, 1986.

Olson, Steve. *Biotechnology*. Washington, D.C.: National Academy Press, 1986.

Pagels, Heinz R. *The Cosmic Code*. New York: Simon & Schuster, 1982.

Palais, Joseph C. *Fiber Optic Communications*. New York: Prentice-Hall, 1988.

Plossl, George W. *Managing in the New World of Manufacturing*. New York: Prentice-Hall, 1991.

Porter, Michael E. *The Competitive Advantage of Nations*. New York: Free Press, 1990.

Rheingold, Howard. *Virtual Reality*. New York: Summit Books, 1991.

Richerson, David W. *Ceramics Applications in Manufacturing*. Dearborn, Mich.: Society of Manufacturing Engineers, 1988.

Sanchez, Diane. "New Business Opportunities," *Success* Magazine, May, 1992.

Schodt, Frederik L. *Inside the Robot Kingdom*. New York: Kodansha International, 1988.

Shromme, Arnold. *The Seven Ability Plan*. Moline, Ill.: Self-Confidence Press, 1989.

Stoll, Clifford. *The Cuckoo's Egg*. New York: Doubleday, 1989.

Strong, Dr. A. Brent. *Fundamentals of Composites Manufacturing*. Society of Manufacturing Engineers, 1989.

Talbot, Michael. *Beyond the Quantum*. Macmillan, 1986.

Thurow, Lester. *Head to Head*. Morrow, 1992.

Vesey, Joseph. "The New Competitors," Unisys Corporation, 1991.

Weiss, Julian M. *The Future of Manufacturing*. Red Bank, N.J.: Bus Fac Publishing, 1987.

Weizer, Norman, et al. *The Arthur D. Little Forecast on Information Technology and Productivity*. Wiley, 1991.

Weller, E. J. *Nontraditional Machining Processes*. Society of Manufacturing Engineers, 1984.

Index

accounting, trends in, 159–160
"ACT!" by Contact Software International, 170
adhesives, advanced, 179
alloys
  dispersion-strengthened, 221
  microduplex, 221
  radiation-resistant, 221
alternative energy production, 125
American Production and Inventory Control Society (APICS), 171
American Renaissance, 20
American Society of Quality Control (ASQC), 11–12
America's New Economy, 113
Andersen Consulting, 165
Apple Computer, 198
architecture, information systems, 238
Arthur D. Little, 85, 166–167, 195, 237
  Forecast on Information Technology and Productivity, 85, 166–167, 195, 237
artificial intelligence
  likely deployment path of, 255
  product-resident, 148
AT&T, 165
A. T. Kearney, Inc., 173, 180
Ausubel, Jesse, 72
automated guided vehicles (AGVs), 96, 144
automated material handling equipment (AMH), 96
automated process planning, 48
automated storage and retrieval systems (AS/RSs), 57, 96, 144
automation, future growth of, 249

bar codes, 74
Barker, Joel, 173
benchmarking, 229
biocomputers, 240–241
Biocomputers, The Next Generation from Japan, 240
biomaterials, 224–225
blanking, chemical, 179
Boeing, 86
bonding and fastening equipment, 94
bonding processes, molecular surface, 179
"Brock Activity Manager Series," 170
Brock Control Systems, Inc., 170
budget deficit, federal, 116
business administration
  machines, 234–241
  manpower, 130–131
  methods, 153–168
business logistics services, 134
Bylinsky, Gene, 126

Cappo, Joe, 117
CATIA, 87
cellular manufacturing, 109, 180
ceramics, 70
  conversion equipment, 91
  fiber-reinforced, 221
  fiber reinforcement of, 190–191
  microcracked, 221
  production methods for, 190–191
  second-phase dispersion of, 190–191
  transformation-toughened, 221
change management, ix
chaos theory, 124
Chrysler, 87
coating equipment, advanced, 246
college-educated manpower in manufacturing, 119
Competitive Edge, Research Priorities in U.S. Manufacturing, 213–214
competitor influences on manufacturing, 22–24, 121–123
composites, 70
  autoclave curing and bonding of, 189
  automated tape lamination of, 188
  bonding and joining of, 190
  ceramic matrix, 221
  cutting, drilling, and machining of, 189
  cutting of uncured, 188–189
  filament winding of, 189
  manual lay-up of, 188
  manufacturing equipment, 92
  matched-die molding of, 189
  metal matrix, 221
  production methods, 187–190
  pultrusion of, 189
  resin transfer molding of, 189
  spray-up methods of producing, 189
  vacuum bagging of, 189

compound document exchange (CDE), 167
computer-aided acquisition and logistics support (CALS), 167
computer-aided design (CAD), 44–46, 86–87, 111, 165
computer-aided manufacturing (CAM), 111, 165
Computer Integrated Enterprise, 207
computer integrated manufacturing (CIM), 10, 167, 171, 180
computer sciences, 165
concurrent design teams, 137–138
concurrent engineering, 111, 171–173, 212
configuration management, data required for, 217
constraint management systems/software, 219
"Contact Ease" by West Ware, Inc., 170
containers, 227
convergence
  sequential, 101, 103–104
  simultaneous, 101–102
coordinate measuring machines (CMMs), 90
cost management, trends in, 159–160
Countdown to the Future: The Manufacturing Engineer in the 21st Century, 180
Crispell, Diane, 118
critical characteristics, 218
critical path
  areas likely to be on the, 260–261
  definition of, 253
  identifying the, 253–261
Crosby, Philip B., 51
Cuckoo's Egg, The, 24
Cummins Engine, 165
customer support
  machines, 250–252
  manpower, 146–149
  materials, 228–233
  methods, 58–60, 197–202
customer support functions, future, 200–201
customer values, changes in, 168–169
cutting, water jet, 189
cycle reduction, 52–53
cycle-to-market, 109

Dassault, 87
data encryption standards (DES), 166
Data General, 87
data security, 165
decentralized factories, 168
decision support systems/software, 219
demand elasticity and forecasting, 105
dematerialization, 71
Deming, W. Edwards, 11, 51
demographics
  of future manufacturing workforce, 118–119, 127
  and manufacturing, 118–119
  of workforce, 26
designer materials, 179
design for logistics (DFL), 133–134, 212
desktop publishing, 60, 86
diagnostic equipment for remote product support, 251
Digital Equipment Corporation, 87, 165, 207
discovering the future, 173
distribution
  channels, 192
  machines, 248–249
  manpower, 140, 144–146
  materials, 225–228
  methods, 55–58, 191–197
  requirements planning (DRP), 56, 145, 194
  strategies, future, 194–197
  tactics, future, 194–197
Drexler, K. Eric, 179
Drucker, Peter, 205

economic influences on manufacturing, 17–20, 115–118
EDS, 165
education and training, national strategy for, 265
electronic data interchange (EDI), 57–58, 96, 195, 242
electronic development fixtures, 67
electronics manufacturing equipment, 91
enabler influences on manufacturing, 24–25, 124–126
Engines of Creation, 179
engraving, chemical, 187
environmental factors affecting manufacturing, 13–25
environmentalism, 22, 111
expert systems in concurrent engineering operations, 173

factory locations, future, 167–168
fading, 255
Federal Express, 59, 134
Feigenbaum, Armand V., 51
fiber, 70
fiber optics, *see* optic fibers, production methods for
Fisher, Anne, 156
fixturing, advances in, 246
flexible manufacturing system (FMS), 111, 180, 246
Food and Drug Administration, U.S., 16
Ford, 87
Forester, Tom, 61, 69
Forward Thinking, 110
Fundamentals of Composites Manufacturing, 188
Future Scope—Success Strategies for the 1990s and Beyond, 117
FUTUREVIEW, 80

gain-sharing, 162
generally accepted accounting principles (GAAP), 64
General Motors, 86
generative process planning, 48–49
Golden, William, 115
government
recommendations for U.S., 264–265
regulations, 114–115
government and industry, relationship between, 113–115
grinding
electrical discharge, 187
electrochemical, 179, 187
electrochemical discharge, 187
orbital, 186
group technology (GT), 46, 67, 180, 212
growth rate, American industry, 10
growth rate analysis, technology, 259

Halberstam, David, 105
Hamrin, Robert, 113
Harris, 87
Harvard Business Review, 205
Head to Head: The Coming Economic Battle Among Japan, Europe, and America, 115, 264
health care costs, impacts on manufacturing, 117
Hewlett-Packard, 87
Honeywell, 87
human resources methods, 160–168

IBM, 165
incursion probability calculation for new technologies, 257
industry, recommendations for American, 262–264
information network management, 236
information processing management methods, 163–167
information systems hardware, costs of, 240
information systems services, outsourcing of, 165
information technology systems return on investment, 234
infrastructure in product and customer support, 200–202, 233
integrated information systems (IIS), 166
intelligent processing equipment, 178
interactive video/audio-based services training, 251
intermetallic compounds, 221
ionic charge, 257

jobs in manufacturing, number of, 117–118, 127
Juran, Joseph M., 51
just-in-time (JIT), 109
manufacturing, 10

Kaminuma, Tsuguchika, 240
KMPG Peat Marwick, 165
knowledge-based systems, 228
Kodak, 165

Lamb, Robert Boyden, 253
Levilink, 85
Levitating Trains & Kamikaze Genes, 20
local area networks (LANs), 89, 244
logistics, 56–57
Luce, Stewart, 188

machines
used in customer support, 97–98
used in defining products and processes, 86–88
used in distribution, 95–97
used in obtaining business, 83–86
used in production, 88–95
used in product support, 97–98
machining
abrasive flow, 179, 186
abrasive jet, 179, 186
electrical discharge, 179, 187
electrochemical, 179, 186
electron beam, 179, 187
laser beam, 179, 187
plasma arc, 179, 187
ultrasonic, 179, 186
Macintosh, 198
Made in America, 16, 47
Malcolm Baldrige award, 12
manpower issues
dominant, 127–129
in manufacturing, 26–32
manufacturing, origins of, 3–4
manufacturing productivity, 7–8
manufacturing resource planning (MRP II), 8–9, 56, 165
market analyses, 242–243
marketing and sales
machines, 241–243
manpower, 131–134
materials used in, 207–210
methods, 40–44, 168–171
Marketing Revolution, The, 208
market share, importance of, 11
material-handling equipment, 249
material requirements for specific industries, 71
material requirements planning (MRP), 8–9, 56
materials
advanced engineered (AEMs), 213
barrier, 228
challenges related to advanced, 247

materials (*continued*)
currently dominant, 61–62
cushioning, 227–228
development systems/software, 219–220
electronic, 222–224
issues, dominant future, 203
magnetic, 223
microprogrammable, 224
photonic/optic, 224–225
shielding, 228
structural, 221–222
used in business administration, 63–64
used in customer support, 75–76
used in defining products and processes, 66–67
used in distribution, 73–75
used in obtaining business, 64–66
used in production, 68–73
Materials Revolution, The, 61
Materials Science and Engineering for the 1990s, 220, 224
Matsumoto, Gen, 240
McDonnell Douglas, 87
"mean time to respond to changing customer demand," 205
metal casting equipment, 90
metal removal equipment, 90–91
metals
and alloys, 69–70
fine-grained, single–phase, 221
metal-treating equipment, 93
metalworking processes, changes in, 186–187
methods, trends in future manufacturing, 153
metropolitan area networks (MANs), 243
microcomputers, 80
microfabrication, 178, 241, 246
milling, chemical, 179
minicomputers, 81
model, Duncan's manufacturing, 1–3, 99–105, 254-261

nanotechnology, 179, 210, 244
National Association of Purchasing Management (NAPM), 171
National Research Council, 118, 168, 220, 224, 245
natural language processors, application of, 247
NCAD, 87
nontraditional cutting and finishing, future of, 246
nontraditional machining processes, 179
Northrop, 87
numerical control, 7, 139

office document architecture-office document interchange format (ODA-ODIF), 167
open systems interconnection (OSI) standards, 166, 195
optical character readers (OCRs), 57
optic fibers, production methods for, 191
optic material production equipment, 93
organizational structures, 112
organization size, changes in, 150–152

packaging
equipment, 246
materials, 74
paper, 70
pay-for-knowledge, 162–163
pay-for-performance, 162
Perkin-Elmer, 87
PIMS Principles, The, 11
plastics, 70
Plastics News, 168
plastics production equipment, 91–92
point-of-sale data collection, 84–85
political influences on manufacturing, 13–15, 113–115
polymers, 222
prime, 87
process integration, 111
process intelligence, 217–218
process management systems/software, 219
procurement, future changes in, 180–183, 212–213
product and process definition
machines, 243–245
manpower, 134–138
materials used in, 210–214
methods, 44–49, 171–180

production
control, future changes in, 179–180
data that will be required for, 212–217
machines, future, 245–248
manpower, 138–140
materials, 214–225
methods, 49–53, 180–185
software that will be required for, 218–220
product life cycle profitability, 129
product support
machines, 250–252
manpower, 146–149
materials, 228–233
methods, 58–60, 197–202
"Profile 21: Issues and Implications," 178
programmable logic controller (PLC), 218
proximity and receptivity factors, 257

quality, 11, 109
changing definition of, 184–185
quality assurance, future changes in, 184–185
quality function deployment (QFD), 46, 141, 185, 213
quantum physics, 24–25

Reckoning, The, 105
research and development, national strategy for, 264–265
retirement benefits, 119
return on investment (ROI), 129
return on net assets (RONA), 129, 140, 160
robotics in distribution, 195
robots, intelligent, 246
Running American Business, 253

sales forecasting techniques, 84
Sanchez, Diane, 170
Saturn , 183
Science and Technology Advice to the President, Congress, and Judiciary, 115
sensor technology, future of, 247
service center management, 230
service industries, 120
sheet metal processing equipment, 90
Shulman, Robert, 208
simulated test marketing (STM), 208–209
simulation
  in product and process definition, 244
  software for production, 218–219
social influences on manufacturing, 20–22, 118–121
Society of Manufacturing Engineers (SME), 173, 180
Sony, 110
Sperry, 87
standards for data exchange, need for, 266
statistical process control (SPC), 109, 184
statistical quality control (SQC), 11
stone, 70
strategic alliances, 23, 111–112, 171
strategic planning methods, 33–36, 153–155
Strategic Technologies for the Army of the Twenty-First Century, 245
Strong, A. Brent, 188
Success magazine, 170
supercomputers, 80
superconducting quantum interference devices (SQUID), 224
superconductors, 223–224
supervision, factory floor, 185
supplier base, global, 182
systems integration, 165
systems standards, cross-functional, 236

Taguchi, Genichi, 51
tax structure impacts on manufacturing, 116–117
Taylor, Frederick, 4
"teaching factory" systems/software, 220
teams, customer advisory, 229
technical workers, demand for, 118
technologies, life cycle of, 4
technology migration, natural paths of, 255
telecommuting, 167, 240
Through the Looking Glass, 252
Thurow, Lester, 116, 264
total quality control (TQC), 11
total quality management systems (TQMS), 11
"Toward a New Era in U.S. Manufacturing," 118, 168, 266
training, customer, 230
transportation equipment, 249

Unbounding the Future, 179
unigraphics, 87
unions, labor, 31–32

valence, 255
value analysis/value engineering (VA/VE), 45
values of future manufacturing workforce, 119, 129
variant process planning, 48–49
Vesey, Joseph, 110
video-based customer training, 250–251
videoconferencing, 82–83
virtual reality, 124–125, 252–253

wage levels, international, 116
Walker, John, 252
Ward's Automotive Yearbook, 72
welding processes, advanced, 179
wide area networks (WANs), 243
Winning the Brain Race, 28
wood, 70
"Workers in 2000," 118
Workforce 2000, 21
World Competitiveness Report, 264
World Future Society, 80

Yalow, Rosalyn, 115